THE SAVVY
SOCIAL MEDIA GUIDE

IMPROVE YOUR SOCIAL MEDIA SKILLS...
NOW!

by DR. STUART H. SCHWARTZ

ENDORSEMENT

STUART SCHWARTZ IS SMART, TIMELY AND RIGHT ON TARGET, IN THE SAVVY SOCIAL MEDIA GUIDE.

His insight is invaluable. His writing is real, approachable and quite understandable, demystifying social media and marketing & advertising. He speaks directly to students and graduates (whom he calls the LAC bunch… life after college). His work is perfect for those tired of the "same old" approach to academia. Who else can find a way to connect the likes of Homer Simpson, the Beatles and even the musical *The King and I* to the brave new world of social media? Stuart Schwartz can. He also brings in key research and introduces us to "Strategic Dualism" and "Virtual Relationship Clouds." But the magic comes in drawing on a wide range of real life stories and examples from decades in the business world. Schwartz brings his keen insight, years of personal experience and an uncommon approach, with his remarkable gift of communication. This is Social Media from an entirely different perspective… but practical and with heart! Schwartz is the modern day Solomon of Social Media.

The Savvy Social Media Guide is an indispensable guidebook to Social Media in the world of marketing. This book is perfect for our shifting media. It's such a useful and powerful book, I wonder how many more Television Emmy's I could have won if I had Schwartz's insight on Social Media a few years ago. Schwartz' book promises to break the mold, in a good way!

Bruce Kirk is on the communication faculty at Liberty University. He spent more than three decades in the media across the country at ABC, CBS, CNN, NBC and FOX affiliates and is the winner of 5-Television Emmy's.

THE SAVVY SOCIAL MEDIA GUIDE
BY DR. STUART H. SCHWARTZ

ISBN 13: 978-1-935986-53-9

LIBERTY
UNIVERSITY®
Press

Lynchburg, VA

AUTHOR'S ACKNOWLEDGEMENTS

I'm blessed to be a professor at Liberty University; a university that emphasizes teaching and results. Liberty professors prepare students for the world outside college and teach them to have an impact in their field of choice. We provide the opportunity for students to prepare themselves to *get stuff done*. Chancellor Jerry Falwell Jr., Vice Chancellor Dr. Ronald Godwin, and Vice Provost Dr. Ronald E. Hawkins are leaders who have dramatically grown the university in quality and quantity, making it an outsized presence in higher education. Savvy leadership—as Homer Simpson would put it, "woo-hoo!"

My wife, Sharon, who has kept our family together through frequent moves and business travel, is a blessing. She uses her spiritual gifts of encouragement and nurturing to improve the lives of all around her and I feel the same way about her as Homer Simpson does about donuts, "Donuts, is there anything they can't do?"

I'm grateful for the support of other faculty at Liberty University who are passionately vocational in nature. Specifically Bruce Kirk, now a colleague after a 35-year career in radio and television. He won five television Emmys and is relentlessly practical in his approach to education. In our little corner of the universe, he is leading the way in preparing students for the digital world in which they'll have to compete. I am also grateful to Todd Smith and Ed Edman, graduates of the Larry the Cable Guy University of "Git'R Done." And a shout-out to our new Dean of the School of Communication at Liberty, Norm Mintle, another "Git'R Done" kind of guy.

PUBLISHER'S ACKNOWLEDGEMENTS

A special thank you to all the individuals who assisted in the creation of this publication:

PROOFREADER: Whitney Delaney

EDITORIAL ASSISTANT: Arielle Bielicki

PROJECT EDITOR: Brittany Meng

EDITORIAL MANAGER: Sarah Funderburke

LAYOUT AND GRAPHICS: Carrie Bell

TABLE OF CONTENTS

PREFACE

Blext. Blext? BLEXT! Yes, this is a "blext," the name I came up with to describe this combination of blog and textbook. This book is a different breed of animal, one that blends the personal with a more traditional "texty" approach. Yes, it covers what you need to know about social media marketing, but in a personal, relevant, and digital way. It's not your usual academic snoozer. This book provides a framework for using social media as the foundational strategy for corporate growth in the digital era, all the while emphasizing– drum roll, please – extraordinary and satisfying make-stuff-happen results!

I want to help you develop an approach to social marketing that—during "Life After College"(LAC) or when you're out there in the real world trying to get "stuff" going—has everyone in your company looking at you and saying, "Wow, if savvy had a face, it would look like [insert your name here]!" You will be known for your common sense, your ability to shrewdly apply knowledge in a way that benefits you, your co-workers, and your organization, that's savvy. You'll get stuff done, good stuff, the kind of stuff that produces extraordinary growth for your organization and a smack-your-lips-in-satisfaction career. Through heaping helpings of strategic social communication campaigns, you'll bring customers or donors through the door, make them smile, and encourage them to buy, donate, or join. Your customers will be happy; your company will be happy; and you will be happy. Listen to me and you'll never reach the point of frustration, where you say as Homer Simpson did, "That's it! You people have stood in my way long enough. I'm going to clown college!"

I've been there, done that, getting stuff done in the marketplace that is, not attending clown college. This book is written from the standpoint that you, dear Reader, are most likely a college student or at the beginning your career and interested in what most of us are interested in: "What about me, my career, and the future?" It's perfectly natural. This is my second career, having spent a quarter century as an executive with a variety of media and consumer merchandising companies, I teach and write with a point-of-view crafted in a world in which enlightened self-interest is valued. I assume that you want to win, that you want your organization to win, that you want to grow, and do more, and be more, and enjoy more.

Well, never fear, this blext is here to provide you with an understanding of the principles and practices of social marketing and emphasize its use in growing an organization through the back-and-forth of conversational engagement with customers. We focus on acquiring customers and growing their relationship with an organization as expressed through increased sales, donations, memberships, etc. The individual who creates effective business growth strategies, creates more customers, is an employee to be prized and rewarded for the value he or she adds to the organization.

I want to add value to you as you pursue a career. This blext was written for students and young professionals, those beginning or about to begin their career, who can benefit from the use of the extraordinarily powerful customer growth strategies made possible by the explosion of interactive and social marketing tools in the digital age. I want you to be social media savvy. This brings us to another unusual feature of this blext: The use of "Frequently Asked Questions" (FAQ), the online world's way of quickly providing answers to the questions that pop up most in the minds of users. Now, for our first FAQ:

Savvy involves the practical understanding of something, knowing how to apply knowledge in such a way as to get results. To be savvy is to have common sense, the ability to do stuff in a beneficial way for you and others around you. Savvy is the opposite of "academic," in that, the savvy individual knows how to approach and solve problems. The academic individual, by contrast, can rattle off fifteen theories of dating found in indigenous South American cultures, but can't figure out how to ask that new guy in the apartment building to join her for a cup of coffee. In "Savvy-World" we simply say, "Would you care to join me for a cup of coffee at Starbucks?"

A standard textbook is academic; this blext is an anti-textbook concerned with monetizing your life and helping you to become increasingly more valuable to those around you. It will help you DO something. This book is long on application and short on theory. It assumes that you want to do something, that you desire what you do to have value, and that you desire the consequent job opportunities and promotions as a reward for your impact on the individuals and organizations with which you're involved. Reward, opportunity, promotion: to that we can only comment, as did that great philosopher and educator, Homer Simpson, "Woo-hoo!!"

Read this blext, follow its advice, and generate savvy strategies that grow your company while building the perception that you, personally, are an asset to be prized. This book will give you a different kind of education, mentoring you into a winning relationship with your company and co-workers, helping you to build skills and values that will help those around you, to not only survive, but prosper. Forget the traditional textbook approach – the knowledge-for-the-sake-of-knowledge stuff— that's a dead end. That's like being the first person to check-in on Foursquare at a thousand Dunkin' Donuts just to be able to say "I checked-in at a thousand Dunkin' Donuts," so what? It is what you *do* with knowledge that counts. The digital world offers unparalleled opportunities for an organization to grow through conversation with its customers, using social media to achieve extraordinary relationships and, consequently, extraordinary growth. And Homer (Simpson) nods, "Woo-hoo!"

More than two-and-a-half decades in business have helped me appreciate the value of employees, executives, and tools that are creative, reliable, and focused on results. The idea for this blext came to me when I discovered that students are just like employees, both want to succeed. So I've made my classes a place where students can learn how to succeed in Life After College (LAC).

This blext will assist you in developing the strategies and skills necessary to compete and win in a social and digital market. Compete and win. **WOO-HOO!**

CHAPTER 1

WELCOME
TO THE WORLD OF
SOCIAL MEDIA

INTRODUCTION

BAM!! KAPOW!!

Hear that? That's the impact that social media have had on organizations, marketing, merchandising, and communications. It is only fitting that we describe it in terms pulled from a comic book or superhero flick, because that is the AWESOME impact it has had, is having, and will continue to have on businesses. Social media are revolutionizing the marketing and promotion of products, services, and organizations, both internally with employees and externally with customers.

Just a short decade ago, digital interactive media consumed a sliver of the typical marketing budget. Now, however, social media consumes more than 40 percent of the customer relationship expenditures of many commercial and non-profit organizations.[1] Almost 80 percent of consumers decide on commercial and charitable transactions through information and conversation coming from these targeted and personalized digital media channels.[2] Facebook posts, the increasingly ubiquitous tweets, and the flurry of Instagram photos, social media are increasingly shaping the relationship of consumers to the organizations providing them with products and services.

Internally, the story is the same. For example, consumer merchandising icon Procter & Gamble is driving growth through the use of digital interactive tools like Facebook and Pinterest, both inside and outside the organization, allowing its employees to become one with each other and its customers through the social media *force*—to borrow from the science fiction classic, *Star Wars* – with which it has surrounded itself.[3] Organizations are using the digital tools of social media to dramatically expand the size and depth of customer and affiliate pools, entering into a continuous conversation that, at its most effective, binds the organization and its customers. Employers are looking for individuals equipped with strategic and tactical knowledge of this area which, when done right, has the ability to both dramatically grow the organization and boost the value a skilled individual has for an organization.

Social media marketing approaches involve the use of digital technologies on a variety of online platforms to connect conversationally with customers, members, and donors – the individuals that provide an organization with a reason for its existence – engaging them, in such a way as, to increase the strength of the relationship through growing communication, affinity, and understanding. Social media communication is a break with marketing communications of the past, as it requires sensitivity, honesty, and a willingness to be up close and personal in such a way as to spiritually – offering a connectedness once solely the provenance of individuals – bind the individual and the organization.

THE POWER OF DIGITAL CONVERSATION

In this, the digital age, we no longer have to scream, shrug, and accept things that discontent us; rather, we can do something about it.

Let's go back in time. I read about a trendy new eatery in town that a dining magazine (available for free in a rack outside a convenience store) had raved about. The magazine said that wonderful food and an upscale atmosphere created a good-times-to-be-had-by-all experience for diners. Sure, the restaurant created the review for the advertiser-supported magazine, but still, there had to be some truth in the description, right? So I suggested the eatery to my wife as a location for an upcoming lunch. My wife's face lit the room. "Hey, that's a great idea and you're so sweet!" I beamed. "Why thank you, my dear." Whoa—she loves me!

All went according to plan until we arrived at the restaurant. We walked in and were ignored. Servers walked by with barely a glance at us as we stood in the entryway, still optimistic. The dark floors were littered with food and grime—but hey, this was a busy place. A server finally reassured us, "Be back in a minute for you." She was back in more like ten minutes, but we were finally seated. We waited, waited, waited; waited for the menus, waited for the server, waited for our soft drinks, waited for our food, waited for the wrong food to be taken away and the correct order to be brought, waited for our check, and even waited for the credit card to be returned. Afterwards, we trudged to the parking lot, our mood as gray as the threatening February clouds above us.

I had been cheated. I had used my time and hard-earned dollars for a meal that the restaurant appeared uninterested in serving, and provided—at best—a disappointing experience. There was a time when all you could do was grit your teeth and swear, "Never again." But I, as a savvy 21st century consumer, filled with righteous indignation, returned home and wrote about my experience on a social media site specializing in restaurant reviews.

> *Disappointing. Went for a late lunch on a Saturday; the restaurant was only about 20 percent full, yet we waited 40 minutes until meal arrived—which was not worth waiting even five minutes for. Server generally ignored us. The kitchen staff seemed to have better things to do than cook, judging from the occasional leisurely appearances in the dining room. Also wouldn't hurt if they cleaned....*

I wasn't interested in revenge. No, I was only interested in letting the world know about a restaurant that had made it clear: customers are a bother. I clicked and the review went up on the site while simultaneously posting to my Facebook account. Within minutes, I had reactions from around the country. My sister-in-law, three hundred miles away, was quick to reply. Her daughters attend a university in the area served by the restaurant and regularly visit, staying at a local hotel. During each visit they systematically sample restaurants, adding their 'finds' to the list of places to eat the next time around. Her very public reply, "Guess where we're not going to eat next time we're in town!" and her daughter soon chimed in with "Whoa—place to avoid, huh?"

Four hours later and I had received a half-dozen suggestions from Facebook followers for better restaurants within fifteen miles of my home, assorted 'attaboys' and 'you tell'ems' from other friends, and one email letting me know that, on the strength of my post, the offending restaurant was removed from a list of places to hold a church theater cast party. Within a few days, my wife and I had tried a few of the suggested restaurants, one of which quickly became a favorite for pasta, another for gourmet pizza, and another becoming the place for treating our nieces and their friends from school to dinner.

SAVVY MEDIA FOR SAVVY STRATEGISTS

There was a time when all I could have done after a disappointing experience at a restaurant was shrug my shoulders and vow "never again." Perhaps I would tell a co-worker about it the next day or reply to a friend seeking a good place for Italian dining, "Well, I know one place you don't want to go!" The restaurant would lose a pair of diners and, perhaps, one or two others, but experiences, even bad ones, fade and over time I would forget. Diners would continue to try that Italian restaurant and, more often than not, be disappointed.

But this is the age, of social media, of continuing conversations. We have the ability to both instantly know what others think and instantly express our own thoughts. The vast majority of online users are taking advantage of this. Researcher A.C. Nielsen tells us that nearly 80 percent of those online are using blogs and other social media.[4] Some of the conversation is simply clutter that will soon disappear. Savvy individuals understand that this can provide an enormous strategic advantage for both career and organizational growth.

Social Media is the most effective category of strategic tools that has ever powered organizational growth in the marketplace. IBM, for example, has recognized this in its expansion into social media consulting practices that seek to grow its client companies. There is a reason why, in the majority of business-to-consumer (B2C) and business-to-business (B2B) marketing categories, it is surpassing traditional media advertising in the share of marketing communication resources it commands. The reason is that Social Media works!

· · · · · · · · · **FAQ 2:** Is "social media" singular or plural? Do I say "social media are…" or "social media is…?"

> Both are correct. It depends upon how you are using it. "Social Media" is both a field of study, much like marketing, and a means of communication. You don't say marketing are a great tool to use to shape customer perception and buying habits; you say marketing is a great tool. "Social Media" is a name for a set of communication tools that build relationships and use conversations to market products and services and grow an organization. At the same time, social media are digital media channels that facilitate conversations and relationships. Twitter, for example, is a social medium used to quickly communicate what's happening in your life, or the life of an organization, at a given moment. At the same time, Twitter is an important means of driving traffic in the Social Media category. Facebook is a core social medium in the Social Media category. And so, while it's true that you say 'to-may-to' while I say 'to-mah-to,' in the world of interactive media we say both "social media are…" and "Social Media is…"

Life is now imitating the science fiction of the last century. Everyone is connected. Everyone has the ability to evaluate and comment, to give thumbs-up and thumbs-down, and almost everyone does so at one time or another. Individual drops of news combine to form a vast ocean of digital information that includes the "like" button, comments sections attached to blogs, text messaging, the plethora of apps, and online features that encourage conversation and reaction.

More than ever before, it is possible to know what a restaurant is like before you step into it, the percentage of donations a charity heaps upon its executives in bonuses and compensation instead of starving orphans, or that the auto dealership advertising prices that seem too good to be true are, indeed, too good to be true. In the social media age, it is possible to find out if the mass mailer on mortgages that arrived today is legitimate or a scam, or if the "Mom's Secret Recipe" meatloaf special at the local diner is a mixture of canned cat food, ketchup, and tabasco.

On the other hand, the consequences of providing indifferent, sometimes insulting, service in the Social

Media age are rolling in:

- My brother said, "We won't touch that place" when he travels to southwest Virginia with his family to visit with their daughters. The consequences: a meal at this trendy place would average $25-30 per person, and the four of them might spend more than $100 for dinner. They come about once a month during an academic year of nine months. With tip and an extra dessert or two, that puts their potential spending at a restaurant they like at more than $1,000 a year.

- A daughter almost always invites friends, with one or two dining with them about two-thirds of the time they visit. The annual tab for dinners during visits now pushes $1,400, to be spent elsewhere

- But now, their junior year daughter is also on the varsity lacrosse team at this popular and upscale women's academy. So now, when they have their annual banquet, she tells them no, never at this restaurant, my parents say to stay away. And that's enough to push the team into holding the dinner at another restaurant—causing this local business to lose more than $10,000 of potential revenue!

- Other social media sites pick up the review and other diners are encouraged by the review to add their own bad experiences. The result, at least half of the visitors to this small town in southwest Virginia who might be expected to try the restaurant do not, resulting in even more lost sales.

- Now poor service and bad food has consequences beyond a few disappointed diners, possibly more than $50,000 in lost sales over the next year—if the restaurant survives.

This story illustrates the power of social media. All of a sudden, the marketing catchphrase, "What Happens in Vegas Stays in Vegas" doesn't work anymore. The new world of digital social media is bold and bright on the horizon of the operations of both commercial and nonprofit organizations, and more of the everyday relationships of individuals and companies are on display for all to see.

There are two primary reasons for the emerging power of social media:

1. Social media are now part of the fabric of modern life, both private and public. More than a quarter of the population, of the United States, report that their purchase decisions are influenced by social media. More than a million consumers, a week, are viewing customer service tweets and more than two-thirds of the online users (depending upon what one includes in the category: some researchers claim more than 80 percent, some almost 90 percent) at any given moment are engaged in networking through social media. Of these, more than three-quarters are discussing past, present, or future purchases or plans to buy from, donate to, and/or join organizations. The individual, the savvy individual, who can harness and apply the power of these relational media to grow the customer base of the organization that he or she is part of becomes a valuable organizational asset with a remarkably bright future.

2. Social media, by definition, create engagement through conversation. Social media tools such as Facebook, Twitter, and Four-Square are just a few of the multitude of digital applications that are connecting individuals to larger organizational and special-interest communities to talk about the large and small activities of their lives. Online doesn't just mean online, it means talking and forming affinities for a variety of individuals and organizations. Digital conversation has become the primary shaper of opinion. This conversation, in turn, takes on a life of its own, often growing more widespread and engaging as individuals find themselves in relationships through activity and talk with other members of a community. The community determines the talk and vice versa. For example:

- A community, of Nike shoe "die-hards," talk about the latest products of the company, while a steady stream of newcomers joins the community because of its reputation for knowledgeable insights about athletic shoes.
- The Salvation Army's army of online volunteers contributes to the image of a committed and effective charitable organization. This, in turn, attracts others with a calling for service to its volunteer ranks.
- Enthusiastic customers send a YouTube video, "I love you," to New York City regional grocery delivery firm Fresh Direct and, as a result, bring new customers and brand enthusiasts, further building the image of a company that emphasizes the care it takes to improve grocery shopping for the Fresh Direct community.

NO LONGER ALONE

With the emergence of social media, individuals find that regardless of their location, they are no longer alone. A cast-iron toy aficionado in Iowa can talk and continue a relationship with the same in Florida. A Bali resort specializing in providing surfing and yoga retreats for women can gather like-minded females from around the globe, keeping them together and growing its community through online photo-sharing and conversation. While a rock-climbing club in New Jersey can attract and keep members from up to 200 miles away through memberships that combine on-the-premises climbing with involvement in an online community of like-minded individuals and special offers linked to involvement with the club and the sport of rock-climbing.

United Airlines, much to its chagrin, found that its poor baggage handling techniques and unfriendly customer service – handlers destroyed an obscure musician's guitar by throwing it off the plane onto the tarmac while he watched from his seat on the plane – was the subject of worldwide derision when the musician got his revenge by posting a music video account of the episode. Millions of people viewed the various "United Breaks Guitars" videos on YouTube, forcing United to not only make major changes to its customer service area, but also to add a digital and social marketing presence to shape the online conversations of its customers around the world.[5] No longer did travelers shrug their shoulders at inept service; instead, the responses ignited by the videos shaped a powerful digital community of aggrieved passengers that United now addresses in its marketing and customer service operations.

VIRTUAL RELATIONSHIP CLOUD

The engagement possibilities are endless, depending upon the nature of the community, how active and engaged individuals are within that community, and the totality of the chatter. The conversations surrounding an organization can create a positive or negative experience for customers and potential customers. The goal of the savvy social media strategist is to shape the talk, direct the conversation, and affect the interaction in such a way as to create mutually beneficial relationships between the organizations and individuals brought into affiliation with the organization. The result is a growing bond, an affinity between all involved, and fertile ground for future promotions and/or appeals for the social media strategist who orchestrated this new bond.

Surrounding each organization is a cloud of talk and interaction. Organizations now have the ability through social media to promote affinity ("liking" for the organization) and allow individuals, linked to it, to share and create experiences, all becoming part of the continuing and interactive relationships surrounding the organization. We can think of it as an organizational cloud, which we call a "Virtual Relationship Cloud" (VRC). This virtual relationship environment is much like the digital cloud that Apple or Amazon has created for books, music, and other merchandise consumers.

Amazon members are able to tap into the cloud wherever they go, pulling from it their personalized entertainment and digital activities. Which are shaped by the individual and corporate relationship, and enhanced by digital algorithms (computer programming designed by Amazon that monitors member site viewing and buying habits) that continually interact with community members to create viewing and buying options and recommendations. These, in turn, enhance affinity and lead to loyalty, the result of the organization learning as much as it can about how individual members interact with the world relative to Amazon and its products. The result of these virtual clouds is that communications and promotions from the company grow increasingly relevant to the lives of members and are increasingly welcomed. This increased "liking" then becomes part of the cloud of interactions surrounding the organization. For example, a science fiction lover receives emails detailing the latest movie or book by her favorite author. She is directed to the author's fan page, alerted to the release of a new collectible action figure based on the author's latest book-turned-movie, and a special offer for a book by the author's sister-in-law is sent to the science fiction lover's Kindle digital reader.

The cloud that hovers around Amazon services is made up of three equal parts:

1. The activities and preferences of all of its members.
2. The inclinations, judgments, and actions of individuals (i.e. product assessments and reviews, service evaluations whether complaints or praise, etc.).

The shaping of VRC by the organizations executives through customer service operations, technologies, and marketing communication.

A Virtual Relationship Cloud can strengthen the bond between the individual and the organization. Unlike traditional media, social media grow loyalty through interaction on the part of both the customer and the organization. Customers (members, donors, volunteers, etc.) react simply and naturally once they reach the loyalty stage of a relationship. They do more, give more, attend more, and buy more – a natural and personal reaction– contributing that much more to the virtual cloud of activity surrounding the organization.

The savvy social media strategist deliberately shapes this cloud, which takes on a life of its own as it combines the essence of the activities and branding of the organization with the totality of the behaviors, attitudes, and values of the individuals affiliated with the organization. A student with a strong background in strategic communications and social media becomes valuable to a company by taking the knowledge they have learned in the classroom and applying it to real life competitive situations. The young professional, who uses his or her knowledge of social media to create effective strategies that contribute to the growth of a company, will earn his or her promotion.

He or she carefully chooses the digital tools and technology to strengthen and expand the social media cloud, the VRC, of an organization. Customers and prospects are encouraged to participate in activities that fit both the organization and the types of individuals it attracts. The resulting activity, in turn, grows the organization. You, as a student or young professional, don't just use social media, you use savvy social media. The savvy professional constructs a marketing communications approach that links social media use to the goals of the organization and constructs savvy strategies tied to its target community. Savvy marketing, savvy social media, and savvy application build value, both for the individual and the organization.

A ROSE BY ANY OTHER NAME STILL SMELLS

Social media are also referred to as digital, interactive, conversational, and 1-to-1 media. Call it what you will, social media are rapidly changing the landscape of how organizations inform and persuade their "publics" (a term commonly used in public relations field to refer to an audience or market).

Trade publications report that three-quarters of consumer merchandising companies are "'transforming' their marketing operations to more accurately reflect the needs of the evolving marketplace" for more social transactions and relationships.[6] However it is not just marketing and communication operations being changed by a social media approach; there is not a single part of the modern organization, whether commercial or non-profit, that has not been affected by the ability of social media to shape consumer needs and expectations. Social media provide the instant knowledge and results that have become the norm in most societies. Organizations of all sizes are using the conversation and interaction generated by social media to transform every part of the marketing and operations process. Here are a few of the hundreds of thousands of examples possible:

- "Flash sales—sales that start and end within a short, specific time frame—have become popular with budget-conscious fashionistas who want to snap up designer clothes, accessories, home goods, and even vacations at a fraction of their original price." These sales depend upon social media to quickly spread the word.[7]
- "107-Year-Old Herman Miller Redesigns for 21st Century Customers. The furniture company discovered that selling online starts with learning how to listen. The ensuing customer collaboration has fueled innovation and growth" to push this century-old brand to its best sales in years.[8]
- "Nike, the sportswear giant, is adapting its approach to advertising, agency relationships, and product innovation to reflect the demands of the digital world in each of these spheres. The company has cut TV and print ad spend levels by 40% in the last three years according to industry estimates. However, its overall marketing budget climbed to a record $2.4 billion in 2011as digital and interactive becomes the centerpiece of its marketing efforts."[9]

These are just a few examples of the millions of decisions made daily by organizations informed and shaped by social media. We are at the point where no advertising, marketing, and/or customer service plan would be complete without a social media element. Likewise, there are few public relation programs that do not include social media and corporate human resource departments are hotbeds of digital and interactive communications. Meanwhile, operation executives are using social media to strategically manage and promote their organizations, products, and services. Brand leveraging, management, sales, employee relations, direct marketing, I could go on and on and on, but you get the picture. Social media are…

In summary, Social Media as a marketing category has grown to surpass traditional advertising and promotions in the attention that it gets from both the client and vendor side of the marketing spectrum. Traditional marketing communication trade associations urge promotion professionals to "seize the opportunities presented by social media and capture the attention, emotions, and allegiances of countless stakeholders."[10] Consequently, an ever-increasing share of the advertising and marketing budget is devoted to creating and growing conversational, socially mediated relationships with customers, audiences, and affiliates.

THE TAKE-AWAY

Social media have become the foundation of twenty-first century marketing efforts. The impact of the category on organizational marketing and promotion has been dramatic, quickly becoming the largest and most effective approach to binding both employees and customers to an organization. The traditional tools of marketing communication have been reshaped by consumer expectations in the digital age of increasingly relevant information, interaction, and affinity. However, it requires marketers to continually listen and monitor the "chatter" or "talk," the totality of the digital conversations, regarding an organization.

However, just as the success of an organization in dealing with customers can be instantly pushed into the daily lives of current and potential customers, so too can its failures. Almost all of the top brands in the world now have social media strategies that involve the constant monitoring and shaping of the conversational and interactive 'cloud' surrounding these organizations. Organizations now use digital tools to promote a cloud culture and worldview more uniform and potentially powerful than anything seen in the non-digital past. Social media strategies are now the centerpiece of marketing communication campaigns, and, when properly conceived and executed, are enormously more effective in producing results in the marketplace.

FAQ 3: Why end each chapter with "The Take-Away?" Most textbooks don't do this!

The Take-Away" is key to this book, to successful social media use, and to your life of "doing stuff." A take-away requires you to put yourself in someone else's place: What do you want them to think or do? The impression you're trying to make, the actions you want someone else to take, the thoughts you're trying to get them to think—that's the take-away. The term originates in sales, when you're preparing for a meeting with a client to pitch a service or product. You prepare with a single question in mind: "What do I want my client to take away from my presentation?" If you're meeting with a major donor, you want him or her to take away the impression that your organization is worthy of a major donation, that every dollar given to you makes more of an impact than a dollar sent elsewhere. The take-away is the single most favorable impression and suggestion for action that you can leave with a customer, donor, or supporter.

(ENDNOTES)

1   The SoDA Report: Digital Marketing Outlook (Vol. 2 2012), Retrieved from http://www.slideshare.net/sodaspeaks/the-soda-report-vol-2-2012 (http://www.slideshare.net/sodaspeaks/the-soda-report-vol-2-2012)

2   "The Virtuous Circle: The Role of Search and Social Media in the Purchase Pathway,: white paper, Group M Search, February 2011, http://www.wpp.com/NR/rdonlyres/CA49ED29-06A4-4E10-A1F0-C25BAA35CF2A/0/groupm_search_the_virtuous_circle_feb11.pdf

3   P&g outlines growth drivers. (2011, August 9). Retrieved from http://www.warc.com/LatestNews/News/EmailNews.news?ID=28649&Origin=WARCNewsEmail (http://www.warc.com/LatestNews/News/EmailNews.news?ID=28649&Origin=WARCNewsEmail)

4   http://cn.nielsen.com/documents/Nielsen-Social-Media-Report_FINAL_090911.pdf

5   http://www.telegraph.co.uk/travel/travelnews/5892082/Musician-behind-anti-airline-hit-video-United-Breaks-Guitars-pledges-more-songs.html

6   "Digital Prompts Marketing Shake-up" WARC News, March 1 2012, (http://www.warc.com/LatestNews/News/EmailNews.news?ID=29518&Origin=WARCNewsEmail[3/1/2012 6:39:27 AM]), accessed 3/1/12

7   Cynthia Clark, "Case in Brief: HauteLook's 'Flash Service' Speeds Its Flash Sales," 1to1 Media, (http://www.1to1media.com/view.aspx?DocID=33397), 2/1/2012

8   Eric Krell, "107-Year-Old Herman Miller Redesigns for 21st Century Customers," 1to1 Media (http://www.1to1media.com/view.aspx?DocID=33430), 2/20/2012

9   "Digital Drives Change At Nike," WARC February 12, 2012 (http://www.warc.com/LatestNews/News/EmailNews.news?ID=29445&Origin=WARCNewsEmail)

10   "Take Your Social Media Initiatives to the Next Level," Email from Peppers & Rodgers Group, 3/2012

CHAPTER 2

THE ERA OF HYPER-RELATIONSHIPS

INTRODUCTION

WOW—WHO'D A THUNK IT! We were designed for something more than mere existence. Moss exists. Rocks exist. Cars exist. During class, students exist (barely, depending upon the course and instructor). But unlike cars, rocks, moss, and the occasional student enduring a snoozer of a class, we need something more, something that 'touches' us, makes a connection, awakens us to the real stuff of life. The term *touches* or *touch* is used in some marketing and merchandising industries to describe the point at which an organization makes significant contact with a customer. Underlying the term is the notion that every contact with a customer should be meaningful and add to the organization/affiliate relationship, with every "touch-point" (the intersection between the customer and organization) contributing to the goals of the organization and adding benefits for the individual.

The savvy social media strategist takes a look at this traditional notion of good customer relations and, essentially says "I'll see your *touch-points* and raise you *intensity* and *relationship*!" The result, a hyper-relationship, the keystone of successful social media strategy. The Social Media category takes the ordinary relationships of an organization with its customers (affiliates) and turns up the volume. It kicks it up not just one notch but many, intensifying and extending the relationship out to the lifetime of the customer. It creates a hyper-relationship and a hyper-relationship done right drives growth!

Putting "hyper" into relationship is the core of the approach of the savvy social media marketer. True, in other areas hyper may be thought of as negative. For example, we describe a small child as hyper when he runs around a waiting room, careening from wall to chair to end table, never for a moment sitting still. Dictionaries typically define the word as "excessively active."

FAQ 5: So is hyper is a desirable goal for an organization in its relationships with customers?

> The savvy marketer in the age of social media understands that successful marketing communications looks on hyper as a positive. Individuals cannot get enough of the activity that connects with their core need to have meaningful and beneficial relationships. They now demand what was once thought of as excessive and over the top: the opportunity for 24/7 engagement. Traditional approaches to marketing have elaborate timing plans for campaigns: Start here, stop here, take it up again here, etc.

But social media strategy is different, the focus is on building intense and digital relationships while achieving corporate goals. This leads to a two-pronged approach to strategy in which social media marketing campaigns take place against a backdrop of constant conversation with affiliates, engaging customers and donors in such a way as to promote a continuing and intense conversational relationship. Far from excessive, this continuous relationship is the new normal. We are all *hyper* now!

TWO, TWO, TWO MINTS...UH, STRATEGIES IN ONE!

A popular advertising campaign for a breath mint came up with one of the more memorable advertising tag lines of the sixties: "Two, two, two mints in one." This was introduced in the commercials after characters argued about whether the iconic Certs mint was a candy mint or a breath mint but it was both, two mints in one. In the same way, effective social media strategy is actually two strategies in one. A savvy organization in the age of social media creates and executes focused campaigns to address the challenges of the marketing environment and business needs against the backdrop of continuous organization/affiliate conversation. This leads to 'kicked up' relationships: *hyper-relationships.*

We now understand that individuals do not simply turn off their minds and emotions when not directly engaged with a product, service, or organization. Advertising professionals, especially, had fallen into the trap of thinking of their audiences in terms of episodic campaigns based on what agencies term "flighting," which is a way of scheduling advertising over the time of the campaign to conform to viewer buying patterns. Traditional marketers (again, especially advertising professionals) have developed a worldview in which campaigns begin and end, and advertising, although timed to coincide with the consumption behaviors of the target audience, is contracted for finite amounts of time. Small wonder, then, that the world of traditional marketing tends to be episodic, and traditional integrated marketing communications specialists (how's that for a mouthful?) tend to think largely of bursts of activities that, together, comprise a campaign.

It is rather easy to lose sight of the fact that individuals within a targeted market do not shut down their thoughts when the campaign ends or the television and radio spots hit the allotted 15 or 30 seconds. Rather, David continues thinking about that flashy new car he wants mom and dad to buy for him before he goes away to college. Chelsea talks to her friends about the sorrowful dog she saw on a billboard outside of town, asking their advice, "Hey guys, should I volunteer at the Humane Society shelter, or not?" Brittany is talking with all her friends about that radio spot for the new "college for those who want jobs" long after the commercial has aired and is forgotten. The campaigns stopped, the promotions stopped, and the marketing communications professionals went on to the next challenge, but affected individuals within the market kept on talking and discussing their current thoughts and future plans in relation to the product, service, or organization.

Who they talk with and how much they talk depends upon the reactions of others in what savvy social media marketers call a *Virtual Relationship Cloud (VCR, remember?).* I talked about this before, but I'm reminding you again; after all, repetition is a cornerstone of effective messaging. This virtual cloud is the cumulative effect of all of the interactions and conversations, both negative and positive, surrounding an organization. Within this cloud are a constantly colliding and changing set of conversations surrounding each affiliate. The individual's *circle of significance,* individuals linked by conversation to other individuals, all acting and reacting, supporting or joining, approving or disapproving. At any moment, each affiliate is engaged in relationships with other individuals, jointly shaping opinions and thoughts, rousing to action or inaction, and all points between. These sets of interactions, within sets of interactions, enveloped by a virtual cloud and limited only by the organization and the conversation it evokes, are incredibly complex. It's the stuff of psychology and Sigmund Freud or science fiction and Captain James T. Kirk, depending upon your preferred analogy. But whether you're a sometimes-a-cigar-is-just-a-cigar type, or you're into the beam-me-up-Scottie scene, the conclusion is the same: Relationships are complex, ever-changing, and never-ending. Good, bad, or indifferent, it is what we are all about!

The savvy social marketer creates strategy with human nature in mind. Humanity is designed for engagement, for relationships always building along a continuum toward or away from hyper-relationships. Our lives are often a blur of thought, activity, and conversation, making us ill-suited for fitting neatly into the traditional narratives of advertising and promotional campaigns so beloved by marketing communications folks and the *Mad Men* of the sixties (the hit television drama about Madison Avenue advertising agencies). In the early days of television and radio, broadcasting programs began and ended at allotted times with no consumer-grade recorders for time-shifting. Magazines were circumscribed, with physical distribution by airplane and truck. And out-of-home marketing was consigned to billboards and shopping area signage that had to be physically installed in specific locations.

Marketing professionals took a "now you see it, now you don't" approach. It is easy to see why they developed the habit of thinking of customers (whether business-to-business or business-to-consumers) as episodic in habit, turning off with the television or radio, stopping all thought when the pop-up disappears, going blank once the billboard passes. Quite the opposite: Once a subject, approach, product, or suggested activity strikes a nerve, we begin a process of not only deciding what to do, but using the subject and proposed activities, services, and products as a way to engage other individuals, using the 'stuff' of action and thought (past, present, and/or future) to shape relationships with others.

However, our 21st century marketing worldview and digital reality rests on "the fact that consumers often care about the choices of others" and use persuasive advertising and promotions as a means to reach out, both before and after making a decision and taking action.[11] Connection is constant for all of us at all times! Smartphone applications allow me to connect, from Virginia, with my daughters, who live in Texas and New York City, to let them know that I love them and that they are in my thoughts. My wife goes to visit her mother in West Virginia, but her presence remains in Virginia with me through G-chat and smartphone communication.

Digital interactive media facilitates this constant connection, growing and extending our ability to hold conversations at virtually any time and place about any subject with which we are engaged. You name the occasion, on the street, in the car, while washing the dishes, on trains and planes, over coffee at Starbucks, we are able talk our way through it digitally! This means that just about any activity in life may now be part of creating and extending relationships. My sister, for example, always calls when she is driving alone on a long stretch of a Connecticut parkway after a visit with our mother. There was once a time, in the pre-digital era, when I rarely heard from my sister, we reserved our discussions for family get-togethers. Now she calls about once a week, with an occasional text here and there, usually starting with the obligatory "OMG" and then going on to recount the latest issue relating to our mother, drawing me closer to both her and my mother.

Digital tools are merely a means to extend and intensify relationships, bringing more people, with greater intensity, into our personal and virtual cloud of relationships more often. Now, organizations are people too! They are writ large, bold, and capable of relationship, in a way that previously was only accessible to individuals on a personal level. The 21st century is the era of extended and extensive relationships through social media, providing organizations with the opportunity to know as much, if not more, about its affiliates than many individuals know about their family and friends. Knowledge, when applied correctly, as General Electric put it,

"Brings Good Things to Life." Sale leads provide a great example. HubSpot, a social media technology provider, studied 4,000 business using common social media applications. They found that those which[12]

- Blogged frequently as a marketing strategy got twice as many leads.
- Actively added pages to their websites had nine times more traffic than those with passive sites
- Created more interactive landing pages had up to ten times more sales leads
- Regularly used Twitter had up to ten times the leads
- Readily updated Facebook had more than twenty times the leads

Organizations now have the opportunity to become an integral part of the life of an individual, while that life is lived, not just when he or she is viewing a commercial aired a pre-determined number of times (depending upon the media buy) and one show at a time predetermined by a small circle of network executives. In this, the digital media age, exploiting that opportunity involves kicking it up—deliberately and intensely—at least a notch (preferably, several notches) fostering relationships between the organization and individuals in order to achieve hyper-relationships. This chapter outlines a process by which the savvy marketer can develop hyper-relationships, while at the same time strategically building and managing a more loyal and productive base of customers, donors, and members, it's "two, two, two mints in one."

STRATEGIC DUALISM—COMING SOON TO A THEATER NEAR YOU

The savvy social media user develops hyper-relationships using a two-part approach—two, two, two strategies in one! This is also known as *strategic dualism*, which has the advantage of both sounding intellectual and aptly describing a marketing approach that assumes two strategies work together synergistically (Gesundheit!). The two parts include a *foreground strategy* and a *background strategy*. These strategies run concurrently; taken together and individually, these approaches are critical to the creation of hyper-relationships as well as achieving the immediate goals generated by an organization as it seeks to meet the challenges of its immediate business environment.

FAQ 6: This is starting to get confusing—what exactly is a social media strategy?
Keeping in mind the need to develop and maintain hyper-relationships, **a social media strategy is a two-part plan of coordinated relationship marketing activities through which an organization increasingly engages its affiliates while working toward an established growth goal.** A specific goal dictates a specific course of action in the foreground, all the while building conversation and relationship in the background. Working both parts of the strategy together exponentially intensifies the relationship between the organization and its affiliated customers, donors, and members. It takes a bit to understand, but hey, as someone (not Homer Simpson) once said, "I didn't promise you a rose garden!"

Let's look at an example, a church's membership development strategy, to bring more families into its building, might include the creation of new children's play groups, parent coffee socials, and installing new playground equipment. These activities (and the corresponding promotions and publicity) take place against a backdrop, a strategic background, of continuing socially networked conversation about the church and its affiliates. The center will use appropriate digital tools such as Facebook and Pinterest to enfold specific events and promotions related to the foreground strategy. The foreground strategy involves capturing a young-families-with-small-children segment of the market, while the background strategy involves those same individuals and other targeted segments of the market in socially networked conversations about the organization and their lives.

A complete Social Media strategy is concerned with the nature of the messages sent, the types of interactivity implemented, the *stuff* of the present objectives (let's send a donor email out pronto—we need another $2.1 million for the children's facility), and the conversations. The background conversation invariably involves the specifics of the current foreground strategy and both portions of the strategy are worked by the organization simultaneously. The background is the continuing conversation against which proactive and reactive campaigns are run in the foreground. Google, for example, uses its background (or "spine" as it is termed by the digital giant) to secure "great love" for consumers by making sure the conversation revolves around Google products that "are integral to their (Google users) everyday lives."[13]

"Great love" is another way of describing the creation of a *hyper-relationship* by Google with its customers. The virtual conversation cloud being shaped by the organization is deliberately engaging, reaching out, and involving customers in an emotional hyper-relationship that is, in large part, measured by the intensity of involvement and emotional connection with the brand by Google customers. The company's chief executive officer, Larry Page, puts it this way:

> We have always wanted Google to be a company that is deserving of great love. But we recognize (sic) this is an ambitious goal because most large companies are not well-loved, or even seemingly set up with that in mind....We're lucky to have a very direct relationship with our users, which creates a strong incentive for us to do the right thing ... It's easy for users to go elsewhere because our competition is only a click away.[14]

The background portion of the strategy deliberately seeks to build a relationship of "great love" that goes two ways between customers and the organization. The growing awareness of and sensitivity to its customers propels the company into a deeper and greater relationship as the company culture is shaped by a hyper-relationship in which the pull on the organization is as strong as the pull on the customer.

TALK-TALK

"Talk. Talk." This was the title of the song by the one-hit wonder band *The Music Machine*, a sixties group that produced a top twenty pop musical dissertation on the human condition:

Talk talk
Talk talk
Talk talk
Talk talk."[15]

On the surface, this might not seem like much of an insight. After all, everyone talks and so you're tempted to ask, as actress Clara Peller did (on behalf of Wendy's hamburger restaurant chain), "Where's the beef?" In the age of social media, "talk talk" *is* the beef, and the "Talk Talk" lyrics hit the mark. It is what we do, and it is what organizations must do as the backdrop to their communications strategies. In the digital age, the ability to talk has been extended so that an individual can keep up with a growing network of friends and acquaintances from virtually anywhere at any time. Almost all Americans have a cell phone. Look around, and then listen to a PEW research report put out by the research arm of The Pew Charitable Trusts, "Mobile phones have become a near-ubiquitous tool for information seeking and communicating."[16] Ninety-six percent of Americans have used Facebook. One hundred and forty million tweets are sent each day. Eighty-two percent of 18-29 year olds use some form of social networking.[17] Social networking has become part of the fabric of twenty-first century's existence.

That's what we do, from morning until night—it isthe way of the digital world in a time of constant connection. Constant connection means constant talk, for talk is to connection what water is to a river, the latter cannot exist without the former. Again, that's what we do: talk. We talk to others, we talk to God, to ourselves, we talk to our spouses and friends, to the mirror in the morning, to our animals, some of us to our motorcycles (oh, Harley, you beautiful beast of a machine!). It's what we do as we connect in a constant search for meaning.

Social media provides organizations with a way to turn talk into something more involved: conversation. A conversation is what those in relationships have and a continual conversation, deliberately shaped to benefit all parties, evolves into hyper-relationship: the Holy Grail of Social Media marketing. The savvy social media marketer understands that conversational marketing (correctly implemented) can make the conversation between the organization and its affiliates meaningful. The resulting relationship evolves into a hyper-relationship, continuous and affecting, driving thought and talk far beyond an immediate need to purchase, join, or donate something into meaning and, ultimately, significance.

IF YOU BUILD IT, THEY WILL TALK

A movie that has attained iconic status in American popular culture is *Field of Dreams*. This academy award-nominated movie features Kevin Costner as an Iowa farmer facing financial ruin until he hears a voice whispering, "If you build it, they will come." He plows under a cornfield and builds a baseball diamond, all the while ignoring the jeers of friends and family. The film ends with cars crowding the road to the farm, bringing people to watch baseball. "If you build it, they will come," sometimes the pull of an action, an event, a set of circumstances and conditions is simply too powerful to resist. Such is the attraction of a well-developed organizational social media space on which a strategic background is constructed. The goal of the strategic background is to create a conversational connection that will involve affiliates in a hyper-relationship with the organization, and where episodic promotions and communications in the foreground are energized by the hyper-relationships of the affecting conversations (which move and excite the feelings of affiliates) of the background.

As we noted earlier, individuals want to have relationships, need to have relationships and, in fact, gravitate toward relationships, just like the spectators that jammed the roads to the baseball diamond in the Iowa cornfield in *Field of Dreams*. While you or I may not look like its star, Kevin Costner (I know, I know—speak for yourself, right?), or be able to attract others with the faux-sincerity of the award-winning actor, we *are* able to build a digital capability to facilitate conversation among and between affiliates and our organization. And, to the extent that it is not *faux* but truly sincere, "great love" (in the words of the head of Google) can emerge. This binds both organization and affiliate together. It is against this backdrop that the day-to-day business strategies of the strategic foreground are implemented and synergized.

FAQ 7: So what's all the fuss about "synergy?"

First, check out the dictionary.com for this mouthful: "The interaction of elements that when combined produce a total effect that is greater than the sum of the individual elements, contributions, etc."18 When applied to social marketing, synergy is what happens when two cats fight and the resulting turmoil is greater than either of them can produce on their own. "Synergy" is two students cheating together on an exam, achieving greater scores than either could achieve by taking the exam by themselves (Failing is what happens when they get caught, in this case, the unforeseen consequences of synergy). It is 1+1=3. Synergy results in outcomes that are more than would normally be expected.

The application of synergy to social marketing begins with the understanding that social media have empowered individuals in their relationships with each other and organizations. At the same time, the Social Media category has developed into a set of tools and conceptual approaches that allow an organization to enter into hyper-relationships on a greater-than-the-sum-of-the-individuals scale with correspondingly greater results.

How much greater? It depends upon the products and services, upon the type of business or charity, the category in which the organization competes and, ultimately, the desire by the organization for hyper-relationships with its customers.

Surely every organization would like to have intense relationships! True, but an effective background strategy leading to hyper-relationships is a lot of work, more work than is usual in the traditional marketing and promotion capabilities of your average organization or agency. It takes creativity, systematic implementation, an understanding of the psychology of the affiliate, and a willingness to listen. Not easy, but well worth it. The added work often leads to extraordinary results: more members, more donors, more customers, increased sales, etc.

Successful hyper-relationships with affiliates allow organizations to cut through traditional media perceptions and constraints. This approach strips away the stereotypes of media and popular entertainment; providing an outlet for individuals, within the virtual reality cloud of an organization, to experience true relationship with an organization and each other. In this, the rules and circumstances of personal relationships are writ large; individuals get to experience and participate in building a community through social media. It works the same way as a personal relationship.

Let's go back to our friend, Jennifer. Jennifer has been told that Lindsay is a difficult individual. She is warned, "Someone who values peace will stay away from her!" However, the two often take the same commuter train into New York City, morning and evening. One day Jennifer is thrown together on the crowded train with Lindsay, she finds that Lindsay is refreshing, engaging, and shares many common interests. By the end of the commute, the women have agreed to meet at Grand Central Station for coffee before boarding the train together for the ride home. As the weeks go by, the bonding continues and they become friends. The mediated perception of Lindsay has been changed by the actual experience of conversational Lindsay. Over time, nature (the urge to connect and the attendant and involving conversation) takes its course and the two becomes friends, finding that they share interests, attitudes, and approaches to life, and the conversation develops a hyper-relationship.

FAQ 8: Okay, I get it, but you're introducing a bunch of new terms like virtual relationship cloud and synergy, why the mumbo-jumbo? What ever happened to keeping it simple?

I like to keep it simple. But sometimes there is nothing as practical as a good theory, however complex. A good theory works when formulated as strategy and applied, increasing the chance of a similar application working again in a similar situation. Dictionary.com defines "theory" as "a coherent group of tested general propositions, commonly regarded as correct, that can be used as principles of explanation and prediction for a class of phenomena."19 The Social Media category is powerful, in large part, because of the theories about the natural forces that drive continuing conversations and the successful propositions (also known as "best practices") that allow organizations to promote, enhance, and direct those conversations to further corporate growth.

Mix all of this together and the result is growth, and growth is good! A growing company in a free market tends to be a healthy company, and a growing organization that uses social media as the foundation of its marketing is more likely to gather the kind of feedback from continuing conversations that generates creative and entrepreneurial strategies. So let's not wait for this book to be made into a film and released on DVD, instead, walk through a summary of the mumbo jumbo:

SOCIAL MEDIA STRATEGY: The conversational approach used, by integrating various interactive digital techniques, in pursuit of a growth objective or goal. A social strategy always involves using a coordinated set of plans to drive conversation toward a defined and continuing result.

STRATEGIC DUALISM: An organization is most effective in building socially mediated hyper-relationships when it takes a two-part approach to marketing strategy; I call this 'strategic dualism. This dual approach to strategy is the key to dominating a 21ˢᵗ century digital marketplace where conversations are ubiquitous and required for driving growth. Every strategy should be executed against a continuing strategic background of conversation. Remember, "Two, two, two mints in one!"

STRATEGIC BACKGROUND/STRATEGIC FOREGROUND: Every effective social strategy involves two approaches that are work simultaneously, one long-term and continuing and the other focusing on the more immediate marketing and business needs of the organization. The former is the strategic background, used by an organization to both shape and understand, through conversation, the perceptions of the marketplace. The latter is the strategic foreground, where campaigns tend to be tied to the immediate needs of the organization. For example, a non-profit that needs to boost fundraising at the end of a quarter or the introduction of a new line of casual women's shoes by an apparel merchandising company.

Regardless of the organization or category, marketing in the digital era is at its most natural, affective and effective, when a two-part social strategy is pursued. Applications of digital technology are naturally shaped to take advantage of the tendency of customers to receive and seek information socially. "Socially" means that the information does not stand alone. It is always filtered through a constantly changing perceptual screen of dominant attitudes, values, and behaviors shaped by others.

Influencing all of this is the work of strategy. Strategy tends to naturally fall into two parts. For those more psychology-minded, think of it as a split personality in which one personality is there, always in the background, while the other personality does *The Three Faces of Eve* [20] thing in the foreground (Eve is the movie classic about a woman with a multiple personality disorder). Google, for example, has adopted a "dual social media model" that uses a "spine" of integrated services to support a "social network" of amenities.[21]

THE CONVERSATION CONTINUES

The conversation continues—what a wonderful way to summarize the activities leading to a hyper-relationship. Individuals continue conversation in which they share interests, goals, and the everyday stuff of life in relation to another. A purposively constructed social media background involves individuals in conversation related to the organization, giving individuals the opportunity to relate and form an attachment to various portions of the organization. This follows a very simple fact of human existence: The more interactive we are, the more involved we become, and the more we tend to like things.

That interaction grows into attachment has been a staple of consumer marketing research. A Journal of Marketing Research study, for example, found that the more areas in which a consumer is involved with a product, the greater the intensity of connection. The researchers hitched up their khakis, reached for a metaphorical pipe, and somehow came to a surprisingly realistic (although wordy) insight: The more varied the involvement of a consumer with a product or service, the more likely the consumer is become emotionally involved, talk about it, and recommend it to others. Involvement, they concluded, "exerts a strong influence on consumers' decision processes and information search." [22] They then recommended that consumer behavior experts "stop thinking in terms of single indicators of the involvement level" and instead use an "involvement profile," a set of hyper-relationships"[24]

Hyper-relationships are built through constant, relevant, and effective conversation. The savvy social media marketer has a major advantage that can't be overstated: We seek hyper-relationships as a matter of strategy, thereby strengthening a bond to an organization and its products and services. In addition, all savvy social media strategy involves the use of foreground and background strategies. This is different than traditional marketing communications, in which a goal is set, strategy and campaign are developed and implemented, and then the organization sits back and asks, "How did we do?" In social media marketing, there is no sitting back, as a campaign is never over. There is always conversation, always someone saying something about someone or something, always reactions and activity to gauge. This strategic approach is anything but traditional, playing out against a modern communication background that necessitates the need for a more nurturing and longer-term strategy, and continually involving a customer in a conversation about the organization, its products and services, and its brands.

IN HYPER-RELATIONSHIPS WE TRUST

The goal of the savvy social media marketer is to develop hyper-relationships with a growing percentage of affiliates.

Savvy social marketing understands that a small percentage of customers (nurtured into a hyper-relationship) provide an organization with its greatest amount of activity relative to other affiliates who are less invested, both in buying and evangelizing, that is, advocating the organization to others. An informal rule applies here, the "80/20" rule, which says that 20 percent of a company's customers provide it with 80 percent of its business. When you think about it, this is a good assessment of human behavior. It works the same way in most activities: in sales, a small number of accounts do the largest amount of buying; in churches, a small number of members perform the most service; and in the non-profit world, a small number of donors provide a proportionally larger amount of funding.

The savvy social marketer understands that some individuals are more valuable to his or her organization that those with more value deserve further relationship and the best relationships are hyper-relationships, in which the organization is an emotionally valued friend with whom an individual shares his or her genuine feelings, spends time, and interacts. This need for hyper-relationships has strategic and tactical implications. For example, when choosing social media tools and measuring devices, we must keep in mind the need to go beyond simple emotion and engage our 'best' customer. It is important to differentiate here between, say, a Facebook "like" and the engaged "liking" that powers increased consumer attachment to a brand or product. Customers who truly "like" you in consumer behavior show it by buying more, while clicking on "like" is a simple action that may or may not lead to a hyper-relationship (although it's a good first step). Luxury goods giant LVMH Moët Hennessy-Louis Vuitton S.A. whose portfolio of prestigious brands includes Louis Vuitton and Donna Karan, is careful to measure intensity of interaction and not just the number of "likes" on its Facebook product sites. Thomas Romieu, group digital director for LVMH describes the hyper-relationship Facebook approach of the luxury merchandiser in this way:

> …given "the very strong engagement of people on the Internet generally speaking, it makes sense for brands to want to engage there." **But he said it's not enough to simply gin up a Facebook page and leave it at that (emphasis added).** "Brands have to increase the value of the relationship with their customers through this medium," he said. The company is thus tracking the effectiveness of its brand communication on Facebook using both hard metrics like the number of fans as well as harder-to-measure evaluations such as the level of fan engagement and the evolution of that engagement rate over time. For instance, the engagement rate is measured by adding the number of likes, comments and shares for a brand's page, which is then divided by the number of posts in a given time frame and then divided again by the number of fans. Romieu said the resulting figure will diminish over time, as the brand's popularity rises, because the number of truly engaged fans will be diluted. But, he said, it's important to get the managers of the different brands to agree on that metric, and to establish a benchmark.[24]

The number of fans is important. But so are the volume of comments and sharing activity. Facebook is one of many social interactive media tools they use, taking care to look at the interactive environment they are creating around their brands, the Virtual Relationship Cloud. In the VRC, the frequency and intensity of activities are a measure of the "hyper-ness" of the relationship.

Think of it this way, once again, envision a child who bounces in and around everything in a doctor's waiting room as "hyper." The child bounces off the wall, jumps on the couch, runs through the legs of the scowling lady in the corner, yanks the Bluetooth from the ear of a snarling Yuppie (Young Urban Professional) on the far side of the room, and, for good measure, tears a few pages from the Life magazine that's been on the left corner table in the waiting room since 1969. But in terms of engagement, who is most enjoying the endless wait for the doctor? Not the sour Yuppie, the unhappy lady, or the score of others enduring the tedium of modern medicine's waiting rooms; no, it is the active child who is engaging everything and everyone around him or her that time flies.

THE KING AND I…UH, US

And so, we end with our chapter's beginning: Who'd have thunk it?

It turns out that the foundation of new marketing, of the world of digital and interactive media of the 21st century and scourge of traditional advertising budgets, can be summarized in the words of a classic song from another era. In the 1951 Rodgers and Hammerstein musical, *The King and I*, Anna sings the following song, *Getting to Know You*, as she strikes up a relationship with the children and wives of the King of Siam:

> Getting to know you,
>
> Getting to know all about you.
>
> Getting to like you,
>
> Getting to hope you like me.

This was the beginning of a relationship, a hyper-relationship really, that was to turn increasingly warm and friendly, with a bond that was to become so deep that an entire kingdom was plunged into turmoil by the loyalty it inspired. Ah, relationship.

Now, social media are doing with modern marketing and communication what Anna did to deepen her relationship with king's family, turn the world (in our case, the real world) upside down. Social media are revolutionizing the world of marketing and the promotion of products, services, and organizations, both internally with employees and externally with customers. Both commercial and non-profit organizations are using social media digital tools to dramatically expand the size and depth of its customer and affiliate pools, entering into a continuing conversation that, at its most effective, binds the organization and its customers in hyper-relationships.

If you understand and master this, you will have a critical edge in competing for jobs with organizations that are shifting to interactive media approaches and contending for promotions, the next step up in your career journey. Social media are now part of the fabric of modern life, both private and public. How about some déjà vu? Almost a quarter of the population of the United States tell us their purchase decisions are influenced by social media; more than a million consumers a week are viewing customer service tweets;[25] and more than two-thirds of the online users at any given moment are engaged in networking through social media, and of these, more than three-quarters are discussing past, present, or future purchases or plans to buy from, donate to, and/or join organizations.[26]

The individual who can harness and apply the power of these relational media to grow the customer base of the organization that he or she is part of becomes a valuable organizational asset. We're talking about you, the savvy social media strategist, the guy or gal with a competitive and realistic edge who makes stuff happen and who has the advantage in LAC (Life After College) provided by the value added to his or her career and personal life by a savvy approach to working.

THE TAKE-AWAY

The Holy Grail of social media marketing is, in fact, hyper-relationships with the right kind of people, those with the best 'fit' for the organization. Hyper-relationships, done the savvy way, lead to growth. To achieve this, we take a two-part approach to strategy, constructing a strategic background against and around, which our day-to-day campaigns (strategic foreground) play out. It is so much easier than it sounds, as it is a natural approach that plays to the way consumer perceptions are shaped in an age of socially networked digital media.

Remember, social media campaigns, done savvy, are easy and natural. Hyper-relationships, once developed, take on a life of their own. Why make it hard on yourself? Life is complicated enough! As that awesome scholar and marketing guru Homer Simpson has pointed out, "If something's hard to do, it's not worth doing."

The classic television show, *Mission Impossible*, began with the dialogue, "Your mission, should you decide to accept it…" In the same way, an effective social media campaign starts with the words, "Your mission, should you decide to accept it, is to promote hyper-relationships with the segment(s) of your customer base that will result in growth for your organization and its activities." Homer Simpson, of course, put it more succinctly: "Woo-hoo!"

Love, Ease, Hyper-relationship, Homer Simpson, does it get any better than that?

(ENDNOTES)

11   Persuading Consumers With Social Attitudes, Stefan Buehler and Daniel Halbheer, April 2011, Discussion Paper no. 2011-17, p.2, http://www1.vwa.unisg.ch/RePEc/usg/econwp/EWP-1117.pdf

12   "Lead Generation Lessons From 4,000 Businesses: A Study Based on Real Data From 4,000 Businesses," HubSpot White Paper, February 24, 2012

13   http://www.warc.com/LatestNews/News/EmailNews.news?ID=29680&Origin=WARCNewsEmail

14   http://www.warc.com/LatestNews/News/EmailNews.news?ID=29680&Origin=WARCNewsEmail

15   http://www.metrolyrics.com/talk-talk-lyrics-the-music-machine.html

16   http://pewinternet.org/Reports/2011/Cell-Phones.aspx

17   Esposito, J. (2011). *Social media stats for the C-suite.* Social Media Today. Retrieved From http://socialmediatoday.com/jeffesposito/310611/30-social-media-stats-c-suite

18   http://dictionary.reference.com/browse/Physiology

19   http://dictionary.reference.com/browse/theory

20   http://www.imdb.com/title/tt0051077/

21   http://www.warc.com/LatestNews/News/EmailNews.news?ID=29716&Origin=WARCNewsEmail

22   http://itu.dk/~petermeldgaard/B12/lektion%203/Measuring%20Consumer%20Involvement%20Profiles. pdf Journal of Marketing Research, p. 46, "Measuring Consumer Involvement Profiles," Gilles Laurent and Jean-Noelle Kapferer, Vol. XXII (February 1985), 41-53

23   Laurent and Kapferer, 41-53.

24   http://blogs.wsj.com/cio/2012/04/18/lvmh-managing-its-brands-on-facebook/

25   http://www.banking2020.com/2011/07/05/social-media-statistics-by-the-numbers-july-2011/

26   http://www.powerreviews.com/resources/social-commerce-stats

CHAPTER 3

TOOLS, TALK AND APPS: *Real Uses of* Social Media Tools

INTRODUCTION

WHAT'S FASTER THAN A SPEEDING BULLET, MORE POWERFUL THAN A LOCOMOTIVE, AND ABLE TO LEAP TALL BUILDINGS IN A SINGLE BOUND?

No, it's not Superman; rather it's Social Media, the marketing super-category. Although the 1950's television version of the DC Comics superhero was mighty (I quoted from the dramatic narrative which opened each show) the Superman of the digital era has, alas, been diluted by the modern entertainment elite's preferences for role models with feet of clay and heads of politically correct mush. The Superman of today bleeds, doubts, pouts and is so not digital. Because digital is explosive, digital is new ways to talk and share, individually enabling, and digital is a constant stream of innovation that towers and powers above a traditional media marketplace that had long ago collapsed into a Never-Never land of monopoly and complacency. Every day brings new tools, applications, technology, and new ways to talk and engage, a constant stream of innovation in an industry that continues to exponentially expand the intensity and power of its tools.

So, how do you work this, with so much out there to use and each day bringing so much more *stuff*. Good stuff, technology stuff, applications (apps) stuff, but huge amounts of stuff. Everyone seems to have the same question: how do I know what to do? How to use all this stuff?

It is often difficult to know where to go and what to do with the myriad of tools and techniques showered upon us. When you finally learn to use one app in pursuit of your customers/donors/members (affiliates, remember?), it changes or is replaced by a new and (as its creators explain) dramatically better one. Unfortunately, you were quite content with the old technology; everyone in your organization knew how to use it and now you're going to have to spend time and energy folding in the new or revamped app into your operations, and it just doesn't end.

Well, never fear, this book is here to guide you. This chapter is light on technical jargon but heavy on outlining a framework by which you can choose helpful social tools. It mentions a few specific apps, but generally leaves it up to you to make the choices. After all, there are just too many apps out there. As Mae West, the thirties silver screen sex symbol, said about the opposite sex, "So many men, so little time." That's how I view digital media applications, "So many apps, so little time."

So what's a savvy social media marketer to do? Simple:Prioritize according to what works for you and your organization at a particular time in light of your long-term organizational goals and short-term growth objectives. Yes, this is a bit of management techno-babble, but it's the best way to put it. Translation: do what you're able to do with the resources your company has or can reasonably acquire and know where you're going. Always know where you're going!

Keep in mind why you're here, why you're reading this book: You want to harness the power of social media to grow your company, creating a virtual cloud of hyper-relationships that pillows your company's push for growth *and* personally add value to your organization through your heart for customers. Heart matters! In the mid-nineties, legendary Apple founder Steve Jobs explained why his company and *not* IBM would be the one thriving in the twenty-first century, "they (IBM) really have no feeling in their hearts usually about wanting to help the customers."[1] Feeling in their hearts, what an outlook on the customer/organization relationship! What a wonderful way to grow, utilizing the explosion of social media tools that use the "feeling in their hearts" to power community and conversation and consequently, growth.

We begin this chapter by offering advice on how to organize operations for social presence and corporate growth. This will allow you to effectively manage your operation, at the same time provide your company with a clear view of what the organization is doing to capture market shares, how you have organized to get there and (perhaps best from the individual employee standpoint) answer the question: "How do I, as part of the organization, add value by becoming a focal point for growth?"

We introduce the notion of a *hub-and-spoke* organization of your social media marketing efforts. The *hub* is the digital center point, the virtual place where affiliates go to participate in the richness that is your organization, its products and services. *Hub* applications and sites provide marketing and promotional support for the main activities of an organization. Although we discuss the foundational role of Facebook in strategic social media marketing and mention the value of a few specific social apps as *spokes* (the sites that drive traffic to the hub) we don't survey the available technologies. There are just too many of them, with more arriving daily.

Tools, talk, and digital applications (apps) supply an organization with everything it needs to turn current and potential customers into affiliates who are in hyper-relationship with the organization. The savvy social user focuses on understanding how to structure and use a social network. Once you understand this, fit the technology to your goals during planning. Savvy use of the tools offered by the Social Media category provides organizations with two advantages: (1) Effective social marketing intensely binds affiliates to the organization while (2) generating continuous feedback in the form of conversations that allow organizations to adjust strategies on-the-fly, no longer having to suffer losing strategies while waiting for the market to react.

A social media tool (app) can be evaluated and applied according to a *Dimensions Model*. A dimension is a unifying aspect of an app that describes its functions and advantages. Functions and advantages are never far from our minds in assessing a software application. You have to know how something works and the advantages it offers in order to decide how best to use it, if at all. Tools are, above all, useful. Whether we're talking technology or construction, you use the tool that best accomplish a given task within a specified time and within certain budget parameters. We evaluate and use apps based on the following *dimensions*: their ability to promote *loyalty*, connect to a significant *geography*, *drive* traffic, encourage *gamification* (Wh-a-a-a-t? Patience—I'll explain), provide *reassurance*, support *exchange*, and, through emotional attachment, add value to the products and services of an organization. A company revamps its website, adding an app that makes commenting on customer service easier. Ease of use encourages both more negative and more positive comments, providing a more accurate picture of its relationships with customers, more actionable information, and helping it achieve growth through more realistic strategies to attract customers.

Creating and then nurturing customers is what social media are all about.

APPS: THE FINAL FRONTIER

These are the voyages of the starship Enterprise…

Whoops, got carried away, but Social Media is, in fact, the final frontier of marketing. Around every galaxy there is a universe of new applications. For every problem or issue, there is a likely technology solution on the market or in development. That's just how it is in the digital world, where the Klingons and Romulans are simply two among many breeds of software developers. Apps are developed and introduced every day.

But we view apps in the context of our overriding need for organizational growth. To use a metaphor that would cause an English major to blush (if English majors were capable of embarrassment), we surf the never-ending social media wave because it works, because social media strategies, well executed, are the closest thing to perpetual motion that you can find in the marketing world. As a consequence, our appetite for apps has become insatiable, with consumers and organizations gobbling them up as soon as they are introduced. There are so many, it is hard to choose, which is why we are taking a simple and utilitarian approach: If it satisfies, or promises to satisfy, a customer need and it fits our resource profile (we can afford it, have the staff or can hire the staff capable of using it, etc.) then *use* it.

The savvy corporate social user goes with what works, with what experience and the market tells us is the best approach and both are shouting that connection, hyper-relationship, and clusters of technology linkages propel customer growth. For example, before Twitter and Facebook were established, those of us who ran E-commerce websites used a variety of promotional techniques to drive traffic to our enterprises. However, many of the interactive capabilities powered by online technologies has to be designed and programmed, a lengthy and often costly process. In the social online era, interactive applications have exponentientially more power, simplicity, and ease of use than most of the marketing programming of the past. The result: more interactive relationship tools available to marketers at a fraction of the cost. the E-commerce websites and online networks continue to grow dramatically, their facilitate the buying and selling of products and services over the web facilitated by new application technologies. Non-profit organizations, too, are using new online technologies to strengthen their E-commerce tools and techniques, aggressively expanding their continuing campaigns for donations, memberships, volunteers and a host of other organizational activities.

As head of the direct sales division of a popular shoe and apparel retailer, I took over a moribund web presence and led it to a revenue run-rate equivalent to the total sales of our best performing bricks-and-mortar stores. How did I do it? The same approach we are looking at here:

- We started with the customer service experience and worked our way back: Why does the customer go to our site? She likes and wants to buy our shoes, all of which means that we should make it easier for her to buy our shoes, and all the technology we choose should assist us in pursuing that objective.

- How do we get her to notice us? We used our retails stores located in some of the trendiest metropolitan shopping areas in the United States to create awareness and drive traffic to the sites. We printed our website URL on the bag and shoeboxes, created in-store signage that touted the wide selection available online, made sure that every conventional ad mentioned our online presence, and trained store employees to direct customers to the website when unable to find what they wanted in the store. In addition, employees were commissioned on referrals, making the interest of the company coincide with self-interest. This was an early form of *hub-and-spoke.*

- We did an extensive online advertising campaign, pulling people from relevant online sites with banner and pop-up advertising.
- Marketing partnerships were also a significant force, working with our various partner department stores to direct traffic from their sites to our site and vice versa.
- The website was the heavy-duty customer-pleaser. We fitted our site to our customer, the young, urban female with an attitude and an eye for the unique and trendy, from colors to interactivity, to video, to fashion tips, to the copy that described our merchandise. It received traffic from our stores, promotional materials, emails, and online marketing.
- We constructed a blogging and comment area where we could provide customers with the latest on fashion, using video and live chats to allow customers to interact with our founder and chief executive officer, Steve Madden. Through interaction, we achieved a critical mass of women gushing over shoes and ultimately dramatically expanded our customer base, working both edges of our ideal age demographic.
- We reassured customers with money-back guarantees, a no-questions-asked return policy, and an ordering process that kept customers informed every step of the way.

We used all of this and more to make our website the *hub* of our business, making sure that the experience was exactly the kind they wanted and that it was a place that was exciting, interesting, and in tune with the thoughts, wishes, and dreams of our best customers. That going to our site was (for the young ladies who spent significant amounts of time online and in our stores with us) like immersing yourself after a long day of work in a hot bath filled with bubbles. We told them, "yes, we care, really care," and then showed it by how we treated them, personalizing every interaction.

Personalization is important. Every part of an organization's online presence must be used to further advance the integration of the company with its social network. Every organization has a network of affiliates who are, to varying degrees, involved with it. The goal is to continue advancing these affiliates along a continuum that moves from disinterest, to welcome, to embrace, to I-can't-live-without-you, and, finally, I-can't-wait-to-tell-all-my-friends-about-you. All of this is achieved by matching the appropriate technology with the objectives of the company and constantly working your applications with the desired outcome for the customer in mind. Social technology is merely the means by which the conversational content of the organization is integrated with the conversation of its affiliates.

HUB-AND-SPOKE: NOT JUST FOR SPACE STATIONS

The *hub-and-spoke* approach to organizing the social network of an organization is the most effective way to achieve the desired objectives of a company, putting in place a social operation that allows both immediate and long-term strategies to be worked simultaneously. The *hub* is a central presence that encapsulates the core values and activities of the company, features the core benefits for affiliates of association with the company, and sets the core content and psychology of the conversations. The *spokes* (channeling entities such as Twitter and email marketing-initiated communications, to name two of hundreds of technologies available) drive traffic to the central online presence. This presence can be a social media site (i.e. Facebook) or the organization's central website.

Twitter, for example, in addition to being a place for major-league snarking (what Twitter user hasn't cracked wise and what politician hasn't tried, often disastrously, as clever is not inherent to our political class, the occasional Twitter putdown) has long been known as *the* place to get your enterprise noticed and push traffic to your main site. You have to first define your goals and how you want your market to view you, then comes technology.

Etsy, for example, has used Twitter's digital word-of-mouth to position itself as the little guy's or gal's alternative to corporate Amazon. Etsy is an online handmade goods and craft store collective of sellers that has found a digital sales niche as "a crafty cross between Amazon and eBay," and as a New York Times headline put it, "Where the Craft Babes and D.I.Y.(Do It Yourself) Dudes Are."[2] More users go to Etsy from Twitter than Amazon, with Twitter being,

> [T]he place to tout great Etsy finds and one-of-a-kind items that are both uniquely soul-satisfying for those wondering whatever happened to craftsmanship in this modern world, and trendy for those who value individualism but won't risk falling on the "not" side of the what's-hot-and-what's-not dude/ dudette crowd.[3]

Do you have to use Twitter? No, use what works. A spoke, such as Twitter drives traffic, may not be effective for your organization. Twitter works for Etsy, but may not work for you. A quick online search provides you with multiple online alternatives. Use what works best for your organization and market segment(s). Even the founder of the original photo-sharing companion site for Twitter, Twitpic, has decided to use the strength of his niche companion app to launch a direct competitor to Twitter, Heello. Use what works, this is the savvy mantra of the savvy social media strategist. Look around, investigate the alternatives, experiment a bit, and, above all, keep an eye on what's working for competitors in your category or for the best performers in other categories. If a strategy or tool is working for someone else, well, steal it.

FAQ 9. Hey, hold on! Steal? STEAL!? That's exactly what's wrong with this world, that you'll do anything in pursuit of money and power and, and...

> Whoaaa...now you hold on. "Steal" is a bit of hyperbole—puffing up—to emphasize a point: find out what the best, most savvy operators and marketing folks are doing and then adapt their winning tactics to your organization. It is part of savvy, understanding what works and what does not. Business consultants have a term for it, "Best Practices."

Best practices are tactics that show superior, or better, results consistently. Notice the key words: results, superior, better. You look at the winning techniques and technology, the strategies and organizations that are having success, and you see what you can apply to your organization. This approach works in both the corporate and non-profit worlds. It's one of those awesomely simple "built-in's" for humanity, both on the personal level and social level. If something works over there, then the probability of it working here is higher than not working here. It's part of our operating instructions: You imitate quarterback Tom Brady to learn the timing behind throwing out of a pocket; you go to Bible commentaries to understand how others interpret passages, with the notion that you can build on others' understanding to give you an advantage in your spiritual life; and you look at Amazon to understand how the use of statistical algorithms can assist in generating customer loyalty. Different subjects, but the same process and same humanity apply.

A savvy organization seeks growth, and best practices can assist. A business classic sums it all up in a single title, *Grow or Die: The Unifying Principle of Transformation*. Author George Land points out the simple and obvious result of observing natural and organizational best practices throughout history: Growth is basic, universal, and our best chance at providing an organization with the means to effectively compete and provide transformational experience—what we're now calling hyper-relationship--for both its affiliates and employees.[4] In this century, social networking is the engine that drives transformation.

However, the enormous day-to-day changes in social media technology can be confusing for an organization, which is often confronted by a bewildering array of technology choices and voices screaming, "We're best, we're best!" Simplify your approach by asking a simple question: What works, given the human resource skills I have or can acquire and the type of affiliate/organization relationship I want to build? Then, see what solutions fit, a process that demands a bit of work and a bit of effort. Think about it--does life, which is quite messy, get easier when you concentrate on results? Open your mind and your heart. I have stolen (best practices, remember?) from the approach of one of the greatest professors of all time, Yoda, who advised: "If no mistake you have made, yet losing you are, a different game you should play!"

The technology is grouped into two elements: first, the hub is where perceptions are created, the work of the organization is conducted and conversation shaped; second, traffic must be generated and directed conversations brought into, through, and out of the daily promotional operations of the organization. We call these technological conversation sweepers the *spokes*.

HUB. This is how you organize your technology, regardless of the applications you've chosen. The savvy organization establishes a digital critical mass at a single point on the Web. This major presence is defined by the totality of what the organization does, what it is *trying* to do, and the face it presents to the individuals (customers, donors, vendors, members, etc.) who relate to it as affiliates. No matter how many types of pages and sites an organization has out there (Facebook landing page, Twitter homepage, brand and campaign microsites, Pinterest homepage, corporate website, etc.), a single site, or defined set of sites, should provide a major presence. This major presence is the *hub*, the entity that tells all who go through it, both internally and externally, what your organization is about, what you care about, what affiliates can do with and think about your organization. The hub has two overriding functions: communicate your essence while facilitating the *work*, the stuff of your organization.

Communicate your essence, what do you want people to think about your organization, as evidenced by the totality of conversations passing through and hovering around the *hub*? Fresh Direct, the grocery delivery firm that operates in Manhattan and other metropolitan areas, is renowned for its fresh products, prompt delivery, and courteous employees. Everything about its website *hub* communicates these strengths, from the headlines on its home page ("Daily Freshness Ratings," "Groceries Delivered to Your Door," "100% Satisfaction Guarantee") to the testimonials from customers ("I feel like saluting when I see one of your trucks go by" and ""Fresh Direct is just great. The quality and price of meat, vegetables, and fruits are outstanding").

The hub facilitates the work of the organization. If you're a church, the hub makes it easy to join. If you're a charity, the hub makes it easy to give. In a political organization you need to have an area of your website for contributions, volunteering for campaigns, obtain general information about the candidate; all the *stuff* that must happen in order to make candidates and their campaigns successful. Nordstrom sets the tone (communicates its essence) for a specific type of customer with its stylish graphics and fashionable models who "are all dressed up and everywhere to go" but also provides every opportunity for its customers to buy and shop through its E-commerce functions on the website or directing customers to specific stores.

A hub can be a website, a Facebook page, a blog, or a combination but it must be the major perceptual and functional site for the organization. All other online and offline promotional areas should feed into the hub, from billboards by the interstate, to magazine ads, to Twitter accounts, and beyond. The promotional elements that feed into the *hub* are *spokes.* Think of it as resembling a chariot wheel.

The hub site receives and shapes the conversation passed through the spoke sites. Again, the technology you use for the hub may differ based on the skills and traditions you have available. Consumer merchandising groups that maintain E-commerce websites that sell on a retail level do well to build their social media effort around their central selling site. This is the source of their best customers, the affiliates more likely to buy and talk about the organization and its offerings. Other companies use, say, Facebook as their main site while maintaining an official corporate site for shareholders and investors. The Facebook site is the central organizing force for the organization and its affiliates, with loyalty and marketing programs centering on its pages.

At the same time, savvy social media marketers remain alert for new site possibilities and applications to assist in putting together an effective hub. For example, Pinterest is increasingly being used as a hub for certain types of companies. It has an uncanny ability through graphics and connecting techniques to showcase the lifestyles of the individuals targeted by a brand or product. The American Association of Retired Persons (AARP) uses Pinterest to portray the vigor of its membership, an active set of affiliates who appreciate the positive image they have through the Pinterest account. Again, the emphasis is on what works.

SPOKE. Individuals rely upon digital media to provide guidance in what to think and what to think about. This does not mean that individuals automatically think what you want them to think; there are many factors involved in how an individual sees the world, from family and friends to genetic makeup and upbringing. However, the advantage of the social networking age is that individuals can become involved in conversations that, over time, shape their thoughts and, ultimately, their actions. The involving nature of social media makes for greater commitment once individuals become engaged and affiliate with an organization. The capacity for hyper-relationship may increase the pull of an organization for an individual; this pull is magnified by the individual spokes, which offer opportunities for involvement.

Spokes may or may not be digital. For example, a decidedly non-digital shoebox with a brand website reference graphic is a spoke that offers the shoe consumer an opportunity to further increase affiliation by going to the hub website of the brand. There, he or she may join a frequent-buyer club, opt-in to receive promotional emails, join a "shoe-cruise," respond to an offer to buy a pair of awesome boots with a second at half-price, etc. Interactive-based talk is what keeps all of this going, and there are many ways technology can further the conversation through spokes. A related blog, for example, can offer links to a main site, and the comments on the blog can offer opportunities for conversation; Twitter can push affiliates back to the main site; Facebook is often used to drive clicks through to the hub website.

The spoke portion of the hub-and-spoke is the critical area where conversations are started and continued. It contributes, too, to the notions of Engage and Re-engage. This is where a company can enter into the conversation, making sure to engage those affiliates who are assisting in advancing the goals of the organization. The savvy social marketer monitors the conversation, actively shaping it by responding quickly to negative turns and promoting the positive directions that individuals are taking. You have to have a place, in the digital media age, for people to hear about you. This both initiates and shapes the talk and continuing conversation. Twitter, for example, is a powerful application that allows individuals to get involved with an organization and push traffic to its hub site. One technology blog, Awe.sm, calls it "the perfect storm for referral traffic" which, because of its "openness and the many resulting ways users interact with it" have made it a primary driver of spoke activities.[5]

The primary determinant of a spoke application, that adds value, is whether its structure and nature promotes conversation while driving traffic to the hub and whether it can be used by the organization to increase the volume of positive conversations while quickly responding to negative turns. It should be noted that conversations in the digital age do not have to solely involve text; sharing photos, for example, is conversation, as are audio and video sharing. The content of the sharing, not the media category it represents, is what is important. The savvy marketer must be as prepared to monitor and promote images as well as text, use podcasts to further the work of the organization, and connect to video-sharing sites as needed. The spokes promote involvement, pushing affiliates to engage and spend more time thinking about and conversing with others related to the company and its products. Once engaged, affiliates then become advocates, eventually referring others to the products and services of the organization with which they are in a hyper-relationship.

For the organization, the effect of the hub-and-spoke design of social media technology use is to move individuals within the market along a continuum from simply hearing and thinking about some aspect of an organization to creating *effect*, a connection that will then grow (through the conversation and continual interplay of the hub and spokes) and increase the probability of individuals then telling others about the advantages of affiliation and hyper-relationship.

As intensity increases and the affiliate moves along the relationship continuum, he or she becomes what some label a customer evangelist or advocate for the organization. The hub-and-spoke process involves them in a continuing conversation that produces increasing levels of engagement. Digital relationship experts, Peppers and Rogers Group, put it this way, pointing out the advantages of extraordinarily engaged customers: [6]

> Advocacy goes a step further as a business strategy that places customers' interests ahead of the company's. It is built on trust, and trust is an enduring competitive advantage that pays dividends today and long into the future. In fact, trust has become increasingly important as companies lose control of the brand message to customers who can reach the masses in an anonymous, everlasting way. Advocacy, built on trust, is one of the single most powerful factors in influencing a customer's buying behavior….

Why are advocates so important? They work on a company's behalf to promote the brand, enhance its reputation, and in some cases, drive new business. The goal is to move them up the advocacy involvement scale from mentions through tweets, status updates, and casual word of mouth to influence via blog posts, comments, and direct word of mouth, and then to referral with reviews, blog posts, recommendations, and direct work with the brand as an ambassador.

IN PROCESS WE TRUST

For best results, the choice of tools from the Social Media category relies upon a process of exploration, creation, and implementation. This book will not tell you exactly what apps to buy or what software development tools to use. Instead, think and develop discernment, insight, and penetrating judgment in growing a digital connection to your business. To help, this chapter outlines a process that you will walk through to discover the tools for yourself; ultimately, allowing you to assist your organization to grow both the amount of relationships and the intensity of those relationships possible in this interactive digital age.

Savvy people work their way through challenges, keeping results in view at all times. Results are best produced by using a process to construct a hub-and-spoke social marketing strategy for an organization:

1. **Ask yourself: What am I trying to achieve with my hub?** Sell more shoes, reassure existing customers, implement new ways to donate or volunteer, etc. You get the picture. Hubs are complex areas of online space where the needs of customers (for products and services, for reassurance, for love, loyalty, and ease of ordering) are satisfied. If you're a retailer, then the hub should include a place to order easily, find your bricks-and-mortar stores, check the latest bargains, etc.; in a charity, the hub provides a way to volunteer or donate; in an educational organization, the hub is the center point for current and prospective students and alumni. The hub should reflect the essence of your organization, especially its branding. Does your technology represent the organization appropriately to affiliates during a typical visit to your hub?

2. **What are others in my category, in my competitive space, doing?** Why? Is it working for them? If not, *why* not? For example, Pinterest is the third largest social network in the United States, just behind Facebook and Twitter, but just a few short years ago, in 2009, it did not exist.[7] You go online and research it, trying to understand the competitive and relational advantages the application offers. It doesn't take too long to find that a number of observers feel the social tool offers a significant opportunity when used as a *spoke* application to drive traffic:

> But what makes Pinterest rather special as a social platform is the degree to which it runs counter to the "island" mentality most social networks create…. Pinterest…is driving huge amounts of referral traffic. Relative to its size, it is driving far more outbound traffic than platforms like Twitter or Facebook ever have, making Pinterest of special interest to the enterprise. Unlike Facebook or Twitter, Pinterest feels like it can be monetised (sic) by the enterprise.[8]

Investigate and experiment, it's that simple. Any textbook you read giving a survey of social media is complete for about…oh, three or four nanoseconds but the value of a social media selection process is that you discover for yourself what works and doesn't work for you and your organization, and how best to apply it in relation to your competition, and your organization's need for monetized relationships with affiliates.

How do you do it? Think, savvy reader, think and do. To continue with this example, let's turn to the business-to-business sales category, wherein one organization sells to another rather than directly to consumers. You turn to the many vendors (organizations that sell to other companies) of business services and products, many of which do an excellent job of providing research support that assist clients and prospect organizations in creatively using their products. An example is HubSpot, a premiere social media integration application for businesses. The company suggests, in a valuable e-book, ways that businesses can use Pinterest:

The biggest challenge for B2B companies wanting to use Pinterest as a marketing channel is a lack of visual content. By nature, many B2B companies are selling a product or service in an industry that most likely isn't visual. Here are some ideas of visual content you can post:

Visual content you already have:

Someone at your last company mixer must have grabbed a few photos, right? If so, make a board to showcase your company's culture and pin them. Have executive headshots? Create an "executive management" board and include a bio for each person.

Strong visuals from blog articles

Start using clear, beautiful images in your blog articles with the point of pinning them to your pinboards moving forward. You should be using images in your blog articles anyway! Pin visuals that best highlights your written content.

Infographics & Data Charts

Infographics are all the rage right now, and they are doing very well on Pinterest. If you have any industry data that you can visualize, do so before someone beats you to it.[9]

The common theme is collecting intelligence on the market and competitors. A savvy social organization operates a mini-Central Intelligence Agency, minus the bureaucracy and we-have-ways-to-make-you-talk stuff.

3. **What technology and engagement applications fit with the lifestyles and demographics of the market segments my organization is targeting?** The individuals involved with organizations use social media in different ways. Which social networks do your affiliates use, and how are they using them?

4. **Which technologies integrate seamlessly with existing and planned approaches, and what kinds of resource requirements are needed?** An organization that shapes and manages its social networking requires resources: people to promote and control content, customer service personnel, and outsourced vendors are just a few of the resources an organization must consider in putting together a social media capability. Social networking between an organization and its affiliates can be dramatically effective, but the amount of detailed and often costly effort it takes on the part of the organization is regularly underestimated. Much depends upon your starting point. Does your hub, for example, need to be modified or enhanced? In examining your customers, are their other social network applications—Instagram, for example, the instant photo-sharing application—as useful to your affiliates?

5. **Finally, how will you know when you've succeeded?** By the number of affiliates using a particular application? An increase in sales or donations? More leads for your sales force? Added fans in the seats? By the nature and frequency of conversations? The effectiveness of a strategy is measured by some combination of metrics: the key numbers that combine with continuing customer conversation to tell you where you stand. Every social networking effort should be monitored through counts of critical activity (i.e. click-through rates, sales by time and geography, number of positive mentions, etc.) and conversation feedback. Choosing key statistical, financial, and conversational criteria offers a means of benchmarking your actual results against planned results while comparing your effectiveness to other competitors and application users. Benchmarks provide a scorecard, a way of knowing when success is success, which we'll get into later in more detail.

Simple, really, Benchmarking is setting a standard, laying out a set of goals and checking your progress against that standard or goals. It is something we do all the time in our personal lives but then seem to leave behind in the business world. One of the first lessons I learned as head of marketing and sales for a major media company remains both trite and true: You get what you inspect, not what you expect.

Let's use the example of a mom planning a birthday party for her six-year-old son. Her goal is to hold a successful circus-themed party in her backyard on his birthday. That's the goal, which includes a budget, "A hundred thousand dollars to rent three elephants and their trainers? Are you kidding me? I can rent an elephant costume for Uncle Ned and his boys! Or save on the costumes because they look like elephants anyway!" Now she works her way back: here's when she arranges for a clown; there's when she'll need a load of sawdust delivered; this is when she gets cousin Lulu to hang from a rope in her ballet costume to simulate a trapeze artist; order the birthday cake on this date and invitations on that date, and on and on. Each day, in the six weeks leading up to the party, she checks her list to make sure she is on track, right down to the visits to the therapist to figure out why she ever came up with this ridiculous idea; I mean, why not invite the kids over and plunk'em in front of a television, an all-American and considerably less expensive way. Inspection, simple really, WWPMD (What Would Party Mom Do)?

Your average executive would benefit mightily from the savvy exhibited by Party Mom. A typical marketing effort begins with hoopla and goal-setting …and then, silence. He or she forgets that if they're going to plan for 8% growth this year (in sales, donations, memberships, whatever) they also have to do what Party Mom does instinctively and check the list, make sure they're on schedule and getting the necessary results. In a digitally mediated world, there are so many moving parts that it takes a deliberate and consistent effort to compare the continuing results and activities against a plan that extends from the present to the goal. But, fortunately, that's where a company such as HubSpot comes in.

HubSpot is one of many companies that supply tools for tracking the host of social media initiatives that savvy organizations use to succeed in the interactive digital environment. And HubSpot is dramatically successful (with more than 6,000 companies in 45 countries) for their software and services, ranging from non-profit Christian groups, to hotel operators, to healthcare services, and beyond. The products of the company are remarkable for providing organizations with both a framework for working in a social world, and the means to benchmark and inspect the flow of operations.

HubSpot breaks down social media applications and activities into six categories that provide a framework for savvy social campaigns in all types of organizations: (1) content of social media, which includes blogging and the inbound stream of words coming at an organization; (2) lead generation, after all, nothing happens until someone sells something; (3) lead management, don't just stand there, do something and do it well; (4)

search engine optimization, which pushes organizations to the top of the mind of its market segments; (5) email and automation, emphasizing the weaving of email into all marketing strategies; and last but so not least, (6) marketing analytics, if Party Mom does it, so can you!

Go to the HubSpot website, hubspot.com, and you'll find a host of detailed case studies and white papers that together provide a hands-on and practical guide to both the theory and practice of marketing in the digital age. It is a company dedicated to understanding and inspection, to benchmarking and assisting you in knowing what you're doing, all in service of growth, and organizational growth is what it's all about.

VALUE-ADD DIMENSIONS

Add value, add value, add value, add value, add value!

I'd point out that I'm like a broken record, saying the same thing over and over again, but that is so pre-digital that you may not know that I'm comparing myself to the vinyl record albums of the last century that, when scratched, would keep repeating the music; hence, the broken record metaphor. This is the digital era and, while music distribution has changed, the need for repetition is unchanged. Hence, add value, add value, add value, add value.

It's all about added value, even in the selection, adaptation, and use of applications in achieving growth through social strategies. Before we discuss the dimensions of technology that add value to growth strategies, it's worthwhile to repeat (broken record!) that social applications are ultimately about adding value to both an organization and the consumer experience. In this, technology is not simply technology; rather, it is a means of assisting individuals in engaging with each other and an organization in a relationship that benefits all involved. Adding value. . .

For individuals, social technologies add value to their lives, providing a greater connectedness and meaning through conversation, while organizations use technology to pursue growth through a growing intensity of relationship with greater numbers of affiliates. Savvy is savvy, whether digital or otherwise. The savvy manager goes into a store, looks around, and says: "What kind of experience will I deliver for my customers today? Is there a way, perhaps through signage, by how we treat our shoppers that I can persuade them to stay longer and buy more? What can I do with how I operate, how I run my store to make the customer experience more valued and intense?" The store manager is thereby increasing the value of the retail experience for the customer, who often responds more positively and effectively, as evidenced by the larger role the store plays in his or her life.

This same customer-centric approach is critical to much of the applications and software coming out of the Social Media category. Value-add is a dominant and desired approach. An example of this is Groupon, a deal-of-the-day website that features discounted gift certificates that individuals can redeem at local or national companies. This site takes direct marketing interactivity to the next level, a social level by providing individualized discount programs and customizable rewards that benefit both the organizations, that provide discounted merchandise and services, and their consumers.

Groupon retail customers receive offers and opportunities that fit and *add value* to their lifestyles, while the organizations that use the service as a tool for prospecting are pushed into providing a conversational environment that addresses the preferences and lifestyles of their particular consumers. My New York City daughter and her boyfriend begin each day with a Groupon email offer that perfectly fits their active, take-a-chance lifestyles.

A recent offer to the pair, for example, featured both content and value that provided the stuff of conversation for the next few weeks: "$94 Off Tandem Skydive in East Moriches. Skydiving relies heavily on gravity, like corporate trust-building exercises and home-security systems made from banana peels. Harness the earth's obliging force with this Groupon."

Have fun, reinforce the active nature of leisure, experience something together; the company sponsoring the Groupon and the online environment fits the pair perfectly, combining to add value for consumer targets. Adding value depends upon defining:

> [T]he key reason why your company is the best choice for your ideal customer. It should clearly communicate what your company does, and the unique benefits you have over the competition. Essentially, a value proposition should answer the following questions in a clear, concise manner: 1. What does your company do? 2. Why should buyers who meet your ideal customer profile buy from you, and not one of your competitors?[10]

We add value to an experience in a number of ways or *dimensions*. Those dimensions are:

1. LOYALTY
2. GEOGRAPHY
3. DRIVE
4. GAMIFICATION
5. REASSURANCE
6. EXCHANGE

The dimensions are not mutually exclusive. These characteristics together are used to assess the worth of a social technology for an organization. Social applications should have features that make it simple for affiliates to buy, respond, and/or join. Robust applications have tools and features that can be seamlessly integrated with other technologies used by an organization. Buying or licensing decisions should always be made with the need of an organization for a return on its social media investment. This last point dictates that launches of applications be accompanied by action plans with due dates and specific individual responsibilities spelled out.

Finally, keep in mind that all social media tools must add value to the effect of your social network on affiliates. A tool should not just offer the opportunity for conversation; rather, it should promote interaction that, in turn, promotes the growth goals of the organization and the nature of the target.

Life is not often simple and as it turns out, neither are social networks. Social network strategy may suggest one application but an overbearing boss, for example, may pull a "Dilbert" and insist on another set of applications "because I say so!" What tool do you select? My favorite two words in technology selection, "it depends." In this case, it depends upon your boss, the savvy strategists makes his or her best case and then accepts the decision by authority.

It takes savvy: a practical and shrewd understanding of your market, organization management, culture, and, above all, customers. Following are the dimensions to consider when choosing applications and other technology to involve your affiliates:

DIMENSION 1: LOYALTY.
Social applications should build user loyalty for an organization. Loyalty is near the top of a continuum of emotional intensity. Once a relationship between an individual and an organization intensifies, the individual will tend to choose that organization over competitors and, in fact, will choose to spend time engaged in related activities with that organization rather than with other organizations that don't offer the same engagement. For example, when my daughter leaves town after a visit, my wife and I will drive her back to the airport. On the way I suggest we stop for dinner. "No," she says emphatically, "I want to go to Target to shop, I love Target, and I have coupons on my Target app."

"But you're hungry, remember?"

"Don't care, I love Target and Target loves me." She takes out her iPhone. "Coupons, see? Shopping time!"

"But…" I sputter.

"Shut up," my wife said. "And drive to Target."

Applications that encourage engagement usually promote faithfulness to a single organization and its activities. I give to Salvation Army because it is a charitable organization that helps so many people. However, I am *loyal* to Salvation Army, choosing it over other charities when giving, because during two decades of travel the organization warmed me with cheerful and welcoming bell ringers in cities across the United States, making me feel at home on the road as I walked into various shopping centers. I'm loyal because when I give donations as gifts on the Salvation Army's hub website, I can watch the virtual bell ring in tribute to individual members of my family, on whose behalf I am donating, and triggering loving thoughts. In addition, the site automatically sends emails to my family, letting them know that I made a donation in their name.

Choose applications with the need to develop loyalty in mind. Remember the foundation of our social media approach, TERM. Applications that develop loyalty contribute to the opportunity for conversation (Talk), psychically involve affiliates (Engage), provide for continuing connection (Re-engage), and add value to the relationship for both the organization and the individual (Monetize).

DIMENSION 2: GEOGRAPHY.
We are local creatures, comfortable in the here and now of our places on earth. Geography plays a critical role. Pliny the Elder, the ancient Roman historian who died in the eruption of Mount Vesuvius, is credited with saying "Home is where the heart is."[11] In his case, of course, it is also where the ashes are and that's the point. We are not just *who* we are, but *where* we are and geography often makes a difference. Applications that localize, that bring us to a particular place (such as Foursquare, which is a social application for mobile devices that allows users to 'check-in' at various locations) can increase the intensity of a relationship by adding a geographic dimension.

Many hub applications, such as grocery store websites, provide geographical dimensions through coupons and services for specific stores. At the same time, most website and blog development packages allow companies to emphasize feedback and interactivity by location, but it starts with vision, what would most attract affiliates to your organization and how. Consider, your customers organize the products and services of your company in their own worlds.

A multi-store apparel merchandiser, for example, should have a store locator to help the student on spring break find a store along the interstate to Florida. Meanwhile, a golf shop should help the happy 50-something preparing for a week in Hilton Head, locate the best courses in the new developments off the island for new challenges. What about the couple, with three children, heading off to Dollywood's amusement parks for a week? Shouldn't the Cracker Barrel restaurant chain hub site have a 1-click printable map of all their restaurants along their route? Actually, no, the Dollywood-bound family doesn't want reviews or to see what other diners have said about Cracker Barrel, they know what the brand represents having eaten there many times. Rather, they just want to hop from Deep-Fried Catfish Fillet to Fried Chicken Livers to Homemade Chicken n'Dumplins until they land in Dolly's loving embrace.

Again, it all begins with what you want to do in pursuit of the goals of an organization. Yelp and Urbanspoon are major players in the local restaurant review market, one of the first places diners consult when searching for a restaurants. The functionality is local, tying reader-generated content (reviews) to specific geographies. The behind-the-scenes technology emphasizes geography. Urbanspoon, especially, is an aggressive player that emphasizes seamless backend programming links that allows other companies to use its applications to connect their respective customers to localized reviews; this is known as "multi-platform functionality," a bit of technical jargon that means Urban offers a variety of other companies the choice of including its software in their digital products.

In addition, there are many other players in the restaurant information market. How do you decide which technology products to use? The savvy social strategist begins with something simple: Insights gathered from market research. Research into a market allows organizations to understand the fit of technology to a particular market segment, the uses of applications by consumers, and the utility of the technology and features. Begin by getting online and reading. For example, are you considering the use of Urbanspoon in your social marketing hub? Youare not alone, according to popular technology blogging site TechCrunch:

> The company says its traffic is up by 80%, with mobile growth outpacing the web. The site is now seeing 28 million visits per month, with traffic now split roughly half and half between mobile and web. On the mobile side, Urbanspoon has seen 112% year-over-year growth, while on the web side, it's at 70% growth over last year. Overall, the company saw 255 million visits in 2011, up from 141 million in 2010. Across all mobile platforms, including both apps and mobile web, Urbanspoon is seeing 6 million mobile monthly uniques and 10 million mobile monthly visits. The interesting thing about this data is that Urbanspoon can't always tell when a user hits a particular webpage where that user originated – app or mobile web. That's because many pages within Urbanspoon's native mobile applications are actually mobile webpages built using HTML, a decision that the company tells us has been "great for scalability." (Now to work on improved analytics!). Urbanspoon attributes its growth to several things, from new features launched over the past year, to its continued focus on improving its mobile experiences. Notably, it launched a food diary/check-in feature called Dineline in recent months. It has also been aggressively going after OpenTable with its Rezbook iPad app that allows restaurants to take reservations directly from the Urbanspoon app and website. As of last month, Urbanspoon had over 1,200 restaurants using this service. (It had just 800 in August).[12]

Features and benefits, every application, every piece of software has them. The savvy social user takes the time to understand what's out there, ready and waiting.

DIMENSION 3: Drive. How much drive does the application have? If used as a spoke, does it

drive traffic to your central online website or websites? Does it push prospects and returning existing customers your way? Social networking technology should be evaluated based, in part, on the dimension of *drive*. Pinterest, for example, is developing into a social application scoring highly on the *drive* dimension. Witness this review:

> Any business that relies on driving a high-volume of website traffic to increase sales, should consider joining Pinterest. In fact, early research indicates that Pinterest is more effective at driving traffic compared to other social media sites, even Facebook."[13]

Savvy social users assess the ability of applications to drive traffic and activity. There are a variety of levels to this dimension, but all share in the ability to push individuals to activity, whether it be conversation, buying, information inquiries, or donations. Whatever the defining activity of your organization, there are social applications that can supply this dimension. However, as with all that we've talked about, there are many choices and ways to assemble a social network and the answer depends largely on the characteristics of both your organization and your preferred target audience, in addition to the existing intensity of the relationships between affiliates and the individuals representing your organization.

DIMENSION 4: GAMIFICATION. Okay, here we go with the jargon again! Gamification

is simply an increasingly techno-trendy way of saying, "Let's make sure our social network is easy and fun to use, compelling and encouraging." If you think about it, this makes sense; we do more of the activities we like to do, and less of the stuff we do not enjoy. This applies to just about every part of our lives. I like walking the dog and don't like giving it a bath; so I enthusiastically take the dog out to the park, but find every excuse to avoid bathing it. You eagerly enter a classroom where the professor engages you and makes the subject matter relevant to your life, but avoid the many instructors whose life goals can be described as "all boring, all the time."

This is an increasingly popular term in the software development world and even has its own entry on Wikipedia:

> Gamification is the use of game design techniques, game thinking and game mechanics to enhance non-game contexts. Typically gamification applies to non-game applications and processes, in order to encourage people to adopt them, or to influence how they are used. Gamification works by making technology more engaging, by encouraging users to engage in desired behaviors, by showing a path to mastery and autonomy, by helping to solve problems and not being a distraction, and by taking advantage of humans' psychological predisposition to engage in gaming. The technique can encourage people to perform chores that they ordinarily consider boring, such as completing surveys, shopping, filling out tax forms, or reading web sites. Available data from gamified websites, applications, and processes indicate potential improvements in areas like user engagement, ROI, data quality, timeliness, or learning.[14]

So, let's get trendy and use the trendiest of all words, let's *unpack* it for you. That means, of course, let's break this entry down and turn it into something you can understand, piece by piece. Let's just start with what gamification is, it is simply another way to achieve loyalty by encouraging (through technology and social network design) the continued use of digital tools and networks. There is a whole class of technology and technology development that seeks to create software that makes it simple and desirable for consumers to use. Technology engages when the user experience is enjoyable and productive, just what we want as we put together our organization's social strategies.

The above Wikipedia entry essentially says people like to do what they like to do, and don't like to do what they don't like to do. Therefore, we're better off when we provide social networks that have features that encourage people to use them (access to like-minded individuals, ease of use features, guaranteed satisfaction, secure credit card storage, etc.) and fewer features they don't like (such as complex instructions, boring text, disappointing graphics, tedious routines. So, the message is: encouragement, excitement, and relevance beat discouraging, boring, and irrelevant.

Social networking imitates life. Say you invite someone out for a slice of pizza. You would also like (for the sake of this example) this friend to enjoy the evening and leave with a desire to enjoy your company in the future. So you smile when in the restaurant, take care to show you're enjoying their company, and wear jeans that have been washed at least once in the last quarter. In addition, you've brushed your teeth so that when you breathe across the table, your friend doesn't melt down like the nuclear reactor at Chernobyl. You take care to cover your mouth when you burp so the fumes don't burn off his or her eyelashes. Gamification! You engineer the experience in such a way as to encourage her to enjoy the evening and leave with the desire to see you again!

Choose your applications carefully, thinking of the likelihood of enjoyable experiences by your affiliates. Provide features that allow users to do *stuff* in the easiest ways possible, and then talk about the stuff they do in the most engaging manner possible. Create tactics, design tools, and originate creative that makes the affiliate experience natural, enjoyable, and encouraging. No online versions of burned eyebrows, tedious classrooms, or pointless meetings, gamification.

DIMENSION 5: REASSURANCE. Affiliates, like dogs when you wash them,
get anxious. It's part of the human condition. Anxiety is the natural result of attempting something new, of unfamiliar routines, of buying a product or arranging for service. Will this donation actually be used to feed hungry children or put a hot tube in the executive suite? I don't know this instructor, is she any good? Should I enroll in her class? What happens if I order this diamond ring and it doesn't fit, or she doesn't like it, or my pet gerbil eats it, can I get my money back?

To live is to be anxious. With that in mind, website developers have learned over time that users must be constantly reassured. Credit card payment information, for example, should be safe and secure, desirable not just from a financial point of view but from a human engineering point of view. We need reassurance that our financial information is safe, no one will steal from us, and we'll get what we pay for. Again, the anxiety kicks in: Will I be ripped off; will it be worth it; where's my money going? Money back guarantees are not just a gimmick; they're necessary for all purchases, especially ones that are tied to the hub of a savvy social organization.

There are many types of organizations and commercial categories, but each develops a set of approaches that together reassure customers and win their trust. Social technologies that provide reassurance and build trust are foundational to a savvy social operation. Peppers and Rogers group suggests that trust is built on a foundation that includes these six drivers: empathy, accountability, transparency, customer experience, employee empowerment, employee recognition.[15] We will briefly review these drivers that together promote the trust that is foundational to reassurance

Technologies should be selected and deployed with these trust drivers (features that fuel the social activity and intensity of affiliates) in mind. The customer experience, for example, is significantly affected by applications that are easy to use, do what they're supposed to do, and are relevant to the lifestyle of the customer. Accountability by individuals within an organization is certainly a factor ("Is someone going to make sure I get what I order, treat me properly?" asks the customer), as is transparency. The latter is achieved in part in E-commerce applications by providing personalized space for customers to track and manage their orders, edit account information, and quickly cancel purchases mistakenly made. Employee empowerment, (allowing them to make decisions and work closely with affiliates) adds to trust through more effective relationships with customers and affiliates. It is aided by software that provides information on customers relative to the issues at hand and management routines that emphasize employee decision-making while providing managers with the ability to digitally track and inspect those decisions.

Trust, reassurance and, as we'll see later, love are good ol' fashioned American virtues that, resonate, playing a major role in providing a bright future for the savvy social organization in pursuit of growth. Software features should be developed and applications selected for use by the savvy social user with affiliates in mind.

DIMENSION 6: EXCHANGE. It's *social, social,* dear Reader. All apps and software should facilitate exchange. Exchange of information, purchasing, contracting for services, commenting, all involve interactivity, and interactivity is what makes the digital world go 'round. Interactivity and love, of course, but we'll deal with that later. Now, we're providing guidance to you on choosing technologies, and a key goal of the savvy social user is providing a digital means of involving affiliates. Involvement is based on interaction. The two main areas of interaction are conversation and transactions; the social network you put together should facilitate both. Experts see the demand by customers for interaction to increase in the years ahead. One survey of industry experts regarding customer strategy trends had this to say:

> Customers of every ilk are becoming more social, more demanding, and more vocal. They're also more informed than ever before. Many are enthusiastically embracing the myriad technologies that allow them to more easily communication with and about companies. Customers' behaviors and expectations aren't the only things changing. Interaction channels and data sources continue to proliferation. Companies barely have time to address one before another comes along. These touchpoints offer new opportunities to build customer engagement, but also are fraught with challenges. What's more, any customer experience misstep can reverberate like an echo across a canyon, bringing with it a potential PR nightmare.[16]

Exchange technologies are important, as consumers are now expecting to be able to engage in relationships on a variety of devices. The trends are toward more digital interaction, and the applications you choose should allow seamless integration of devices and applications with the lives of your best customers and affiliates. The following quotation explains rising consumer expectations:

> Consumers took to mobile and social to communicate with brands at increasing rates. This movement clearly represents an evolution in customer engagement fueled by the shift in power from the company to the customer. The increase in customers' control of their experiences will continue to challenge the processes, organizational structures, and systems that companies have put in place to communicate with customers, and will require a change in corporate mind-sets, as well. Companies must adapt to the new ways in which customers want brands to serve them.[17]

So many conversations, so much of the transactions of organizations now happens on mobile devices. The savvy social user should keep this in mind when selecting technologies, taking care to acquire applications that can be migrated to a variety of devices while keeping in mind the increasing share of social time now being spent on those devices.

THE TAKE-AWAY

So many apps, so little time. The savvy social media marketer is one who understands that the growth of technology development is exponential, and that it is close to impossible to understand and apply all of the applications that prolific software developers push to market each day, week, month, and year. Luckily, you don't have to; with a bit of discernment, you too can succeed and success for you, dear Reader, is what this is all about.

The goal of this chapter is to equip you with a point of view that will allow you to assist an organization in producing extraordinary relationships with its affiliates through construction of a strategic social network and, ultimately, growing the organization. You will add value in an area of operations and marketing where the result is often organizational confusion, expense, and the acquisition of little-used and often worthless software.

Savvy social media marketing rests on a foundation of merging the organization's goals with a technology approach designed to maximize the conversations that lead to hyper-relationships. It's all about the conversations. To do this, a hub-and-spoke technology foundation serves as the integrating framework for all applications and social network development. The hub is the central and virtual operations and marketing space for the organization. This is where affiliate perceptions are shaped and re-shaped, and affiliates of the organization get to do the *stuff* of the organization in an encouraging and user-friendly manner.

Do the stuff? That doesn't sound very intellectual! Of course, it's not. It's just simple language, by a simple but savvy marketer (who is also an educator, but don't hold that against me. I really can walk and chew gum at the same time), to describe a simple concept: Every organization needs a central online area where affiliates can go to participate and experience the activities and values that form the core of that organization. That online area can take the form of a corporate website, a microsite, a Facebook page, a blog page, you name it, and the possibilities are endless, with more technology being introduced each day.

Regardless of the technology, the hub advances the goals of the organization. The non-profit, for example, receives more and greater donations through the hub; the centering technology assists the religious organization in signing up more members or promoting greater enrollment in church or synagogue activities; the retailer has a virtual place where its brick-and-mortar customers can go for more and deeper involvement with the company. You get the picture. Keep network construction and feature design that is simple, encouraging, and relevant to the customer.

In applications as in all portions of a savvy social approach, the words of a great philosopher and communications scholar come to mind, Larry the Cable Guy: "Git'r done!"

(ENDNOTES)

1   http://betanews.com/2012/04/27/the-downfall-of-ibm/

2   http://www.nytimes.com/2007/06/24/fashion/24renegade.html?_r=3&adxnnl=1&oref=slogin&ref=style&pagewanted=all&adxnnlx=1335888160-lblY9J8qRHuL2hg91BU0Ww

3   http://www.readwriteweb.com/archives/hitwise_twitter_downstream_traffic.php

4   http://www.amazon.com/Grow-Die-Unifying-Principle-Transformation/dp/0962660515/ref=sr_1_1?s=books&ie=UTF8&qid=1335893857&sr=1-1

5   http://blog.awe.sm/2011/07/14/twitter-drives-4-times-as-much-traffic-as-you-think-it-does/

6   Cultivating Customer Advocates, p. 3 Peppers Rogers group

7   http://mashable.com/2012/04/06/pinterest-number-3-social-network/

8   http://wallblog.co.uk/2012/03/28/pinterests-monetisation-challenges-and-advantages/#ixzz1w9KjDIIA

9   HubSpot, "How to Use Pinterest for Business," P. 23, White Paper, downloaded February 2012 from www.hubspot.com

10   http://www.marketingsherpa.com/article.php?ident=32156#

11   http://quotationsbook.com/quotes/author/5749/

12   http://techcrunch.com/2012/02/07/urbanspoon-traffic-up-80-in-2011-mobile-growth-faster-than-web/

13   Pinterest business, p. 8

14   http://en.wikipedia.org/wiki/Gamification

15   P. 10, http://www.nxtbook.com/nxtbooks/1to1/customerstrategist_2012spring/index.php#/10

16   http://www.1to1media.com/view.aspx?DocID=33302&PreviewMode=full&utm_source=1to1weekly&utm_medium=H&utm_campaign=12262011

17   http://www.1to1media.com/view.aspx?DocID=33321&PreviewMode=promotake place against a backdrop of constant conversation with affiliates, engaging customers and donors in such a way as to promote a continuing and intense conversational relationship. Far from excessive, this continuous relationship is the new normal. We are all hyper now!

CHAPTER 4

CONTENT IS KING
...AND QUEEN AND EMPEROR

INTRODUCTION

CONTENT COUNTS. Content is substance and, in every walk of life, substance is important. Even where substance is rare (politics, for example) it is still paid lip service as a desirable trait, judging by the number of politicians who claim to be other than lightweights. Politics, of course, is where the phrase "all hat and no cattle" is most often applied today, describing individuals who talk a good game but lack substance or *content*. It all comes down to content; there has to be something there and, whether individual information or products, most of us prefer something other than an empty suit. And, in pursuit of corporate growth, content counts.

Effective social media, like the world around us, depends upon the *stuff* of conversations for its lifeblood, and that *stuff* is content. The discount in an offer, the wording of an email, the discussion topic for a comment sections, graphics accompanying a review—all content. To the extent the content is substantive and relevant to the lives and characteristics of affiliates; it has a greater chance of acceptance, affect and involvement by a greater number of people. Again, content. It is a critical part of both organizational and individual existence.

Consider, for example, dating. Despite the attempts of Hollywood to reduce all dating to hook-ups by physically attractive people, most of us prefer a relationship with significance—again, content. We don't want someone who simply looks good but can't carry on a conversation; rather, we want to date people who clean up well *and* can string together at least two or three sentences, ideally not just about themselves. Dating is sex, says Hollywood but they are wrong. Dating is relationship, companionship and ultimately, growing meaning—content! The same applies to the products and services of companies; the individual is the corporate and vice versa. You choose companies based on content; people are promoted based on performance (content); and you fall in love based on content ("I can't wait for you to meet him, he so…so…so fascinating").

Content is king, queen, and emperor. It isczar, czarina and czardines—you get the picture.

Look at it this way. We don't just talk, we talk about something. We don't just drive to granny's house; we stop at an outlet center for shopping and a new restaurant for brick-oven pizza, it's the content of the trip. In social media, we don't just talk, we talk about something: content. We don't tweet, review, complain, or explain nothing; we do so about *stuff* (one of the great social media words) or *content*. Content is a natural and ubiquitous part of our everyday lives, and therefore one of the most significant components of the new world of digital media. Content helps to involve and engage. Content, in large part, determines the attractions and power of social media. When we refer to the backdrop of TERM (Talk, Engage, Re-engage, Monetize), it is all about content. Content is the stuff that interests and thereby engages us; the stuff for continued engagement, exerting a positive attraction; and a way to monetize or add value to the corporate/affiliate relationship.

This chapter emphasizes the need for personalized and strategic content generated by the constituencies of the organization. Conversations have to be about something and that something is content. Content is fertilizer for the corporate lawn, producing growth (SMA—Stretchy Metaphor Alert!!). To do so, a company must continuously monitor the conversations its affiliates are involved in, encourage talk that fits with the goals of the organization, and have a clear vision of what and how it wants the conversations to reflect affiliate/affiliate and affiliate/organization relationships.

In the chapter, we survey and discuss a variety of approaches to building content, with emphasis on the interactive strengths of both user-generated and corporate social media content. Homer Simpson didn't say it, but this is a truth great enough to have come from his mouth: Content is the *stuff* of life, the basis for hyper-relationship, and the common ground of loyalty! Wisdom worthy of television!

CONTENT, CONTENT EVERYWHERE

Content is everywhere. If it seems as though everyone is consuming content; that's because they are! Researcher A.C. Nielsen tells us that three-quarters of Internet users visit a social networking site immediately after going online, while one in four minutes online is spent on a social networking site, including blogs.[1] They go in search of content: Bonnie has a new boyfriend; Jillian is enjoying a cup of hot chocolate in the lodge on the south end of the ski slope; a photo montage of Brittany's vacation trip to Hoboken, etc. We can't help ourselves, it's just what we do as humans; we find stuff to occupy our minds and fill our lives. The result is that content is everywhere and because the digital world is the personal world writ large, it is everywhere by extension. One day on the Internet: 294 billion emails sent, the equivalent of 168 million DVD's of information comes up on monitors, and more than a million hours of video uploaded; the statistics are as endless as the information.[2] What, then, is a self-respecting organization to do, in the content area, with so much to wade through? Well, let not your digital heart be troubled. There is a difference between content and filler.

The great difference between traditional promotions and the new media environment is that so much of content is self-generated by affiliates. In the good ol' days, an organization, to a certain extent, could control the conversation by controlling the content. If you wanted people to think a certain way, there were traditional promotions in which you came up with a catchphrase or tagline. Nike, for example, has effectively used "Just Do It!" to present a daring and no-nonsense mantra to the public, while Campbell's Soup is forever "Mmm, Mmm, Good." The great brands relied upon repetition through such media as television and radio spots, billboards, magazines and newspapers to tell people what to think about their products and services. To the extent they succeeded, consumers were stimulated by commercials and talked about them. Given enough repetition and relevance, consumer merchandisers built a brand, a distinctiveness that set their organizations, products, and services apart from competitors in the minds of consumers in the marketplace.

They rarely knew how effective their efforts were until they saw the sales and heard (usually, through news reports and commentary on Old Media) a general level of support. When they achieved 'brand' status, it was assumed that they had succeeded in getting consumers to talk about them, outside of immediate stimuli, in the form of traditional media advertising and publicity. Then came the digital interactive age, which gave individuals the capacity to talk about an organization and its products and services to greater and growing numbers of people.

No longer was conversation confined to personal encounters with a neighbor, co-worker, friends, and family. Now, the digital world extends conversation and the possibility of significance to growing numbers of people. Talk will grow, the possibility of talk with anyone and everyone, as longs as the talk is about something, about *stuff* that is relevant to the individuals engaging in it. Now, anyone with a digital device talks, as evidenced by more than three billion email accounts and more than half a billion websites, producing a cascade of words and graphics, thoughts and ideas expressed.[3]

Talk we do. We're disappointed or thrilled by a restaurant and we review it on Urbanspoon or TripAdvisor, to name two of the many sites devoted to assessing travel, food, and entertainment. Teenage angst meet Teenage Angst, a blog. Harvard Club, Haitian Expatriates, Lovers of Southern Cozy Mysteries, Swiss Chocolate Lovers, Swiss Mint Chocolate Lovers—there is space for everyone to talk online. There are blogs and websites devoted to just about every pursuit of humanity possible, and all of it presents a continuous blizzard of conversation. Now organizations don't have to wait to see if a brand is developing, if they can achieve distinctiveness; instead, a company can watch its products and services develop through monitoring talk, adjusting on the fly when necessary, and changing strategy based on real-time talk and reaction. Conversations tell us right away whether we are perceived the way we want to be perceived. For example, Mesa Grill is a restaurant owned by the Food Network celebrity chef Bobby Flay, according to its Old Media public relations and advertising, it is a showcase for his culinary skills. That's traditional content. Digital content goes further. Diners are now interactively involved in shaping the market image of a brand, and Mesa Grill diners now have a say in painting the brand portrait of Mesa Grill through comments, on review websites, telling of "service with an attitude and lack of interest."[4]

LET THERE BE TRUTH!

The result: Organizations have realized that by actively shaping the conversations around them (the *content* of their *Virtual Relationship Cloud*) they can modify perceptions and strengthen their image and brand presence through hyper-relationships with their targeted markets. In addition, they have also discovered that, by extending marketing communications to offer socially mediated spaces for conversation, they can provide loyalists with a chance to interact with other loyalists, exponentially increasing the influence of the organization. TurboTax, for example, created a forum for loyal customers whose conversations all centered on the use of the company's tax software. The assumption of this community was that TurboTax was unique, powerful, and useful; by having a forum for continuing conversation, they were able to give advice on use of the software product, strengthening user ties to the organization. Customer conversations have resulted in more than 100 user-generated product updates with this result:

> …the community has not only helped the company to gain real-time insight into what customers' think of its products, but it's also helped to improve customers' likelihood to purchase. In fact, she [corporate communications manager] says that their likelihood to buy jumps 26 percent when they participate in Inner Circle. "I definitely think that because we listen and actively engage with them, that's the biggest motivating factor," she says.[5]

In other words, keeping in touch, actively engaging, and shaping the conversations of customers add value to the products and services of an organization. On the one hand, company engagement with affiliates encourages loyalty, adding to the sense of significance on the part of affiliates ("I matter, I really matter!"). On the other, customer conversations (again, content) that are monitored can yield insights that can often assist in growing a company. This is known as crowd-sourcing. CBS Money Watch explained it this way:

[C]rowdsourcing is a very real and important business idea. Definitions and terms vary, but the basic idea is to tap into the collective intelligence of the public at large to complete business-related tasks that a company would normally either perform itself or outsource to a third-party provider. Yet free labor is only a narrow part of crowdsourcing's appeal. More importantly, it enables managers to expand the size of their talent pool while also gaining deeper insight into what customers really want. With the rise of user-generated media, such as blogs, Wikipedia, MySpace, and YouTube, it's clear that traditional distinctions between producers and consumers are becoming blurry. It's no longer fanciful to speak of the marketplace as having a "collective intelligence"—today that knowledge, passion, creativity, and insight are accessible for all to see…. With a deft touch and a clear set of objectives, quite literally thousands of people can and want to help your business. From designing ad campaigns to vetting new product ideas to solving difficult R&D problems, chances are that people outside your company walls can help you perform better in the marketplace; they become one more resource you can use to get work done. In return, most participants simply want some personal recognition, a sense of community, or at most, a financial incentive.[6]

A company monitors, collects, and interprets the conversations of its affiliates, paying special attention to those in hyper-relationship with it. Then it applies the subsequent insights to its marketing, products, and services to create more value. In addition, content offerings have become so valuable to savvy social organizations that many actively seek insights about particular products and tools from its affiliates. Harley Davidson, for example, introduced an effective advertising campaign by asking affiliates to submit ads and creative briefs.[7] The resulting campaign contained content generated by this carefully constructed appeal to the target segments. The subsequent campaign was so focused and engaged that their target segment became an active partner in content and brand-building, quickly moving from engagement, to loyalty, to advocacy. The campaign delivered "incredible results for Harley Davidson dealers across the US."[8] The campaign worked because an engaged pool of affiliates, built through the application of savvy TERM principles and techniques, became powerful advocates for the brand. Crowdsourcing.org, a trade website for vendors in crowdsourcing promotional techniques, outlined the advantages of having built an engaged, and unashamedly loving, pool of affiliates willing to generate the kind of content that continues to drive (pun intended) the Harley legend:

- They have a very large group of fans
- Their fans are truly devoted to the brand, to the point of tattooing the logo on their bodies
- They care about what Harley Davidson stands for and don't want the company to water down their brand
- They understand their own experiences. Although many of the people who work at Harley Davidson are riders, and they obviously understand their audience, they can't possibly know everything about what their riders experience.
- They already have many brand evangelists who will gladly tell you why a Harley is the best bike in the world.[9]

Harley Davidson continues to use social and interactive marketing to create one of the most involved fan-base in the consumer merchandising world. These affiliates continually show their involvement through their active demonstrations of affection for the brand.

Affiliates play a critical role in the marketing and operations of a savvy social organization. The customers of a company frequently experience the reality (the "truth") of its offerings on a more personal and honest level than most of the employees of an organization. A casual shopper in mall store, for example, will usually be treated quite differently (maybe worse) than a corporate executive entering the same store.

"That's the vice president of marketing, smile!" instructs the store manager. "And pretend you don't know who she is; stay with her, and be helpful."

"Oh, another customer?" says the store manager. "We'll get to him after we've got everyone's lunch orders."

The assumption here is that truth about an organization, as reflected in its conversations and its relationship to the market through its products and services, can enable companies to trim counter-productive marketing, promotion, and operating approaches while introducing more market-friendly products. Crowd-sourcing to generate content offers access to truth for savvy social users because

> [T]he voice of the crowd really does discover and converge on the truth.
>
> In the business world that means that the social crowd will figure out if a product has poor quality, a company is lousy at customer service, or a listed price is different from the average. This new world of information makes it essential that brands, product positioning, and marketing campaigns strive toward truth and authenticity. You're not going to pull the wool over the crowd's eyes, so don't even bother trying. Instead, embrace authenticity. In the same way that "the crowd" has an uncanny way of spotting lies and half-truths, the crowd can also identify if you're truly being genuine. And when they do, that positive word will get around just as rapidly and powerfully as all the negative discussion.[10]

It is important to view content as the source of valuable feedback, which a digital-era organization needs to grow. Content is everywhere; it will continue with or without you, as many organizations have discovered while trying to put an end to discussions (rumored or actual) of poor service, declining business, affronted customers, shoddy products, etc. But they can't because access to great amounts of individuals is no longer the sole domain of companies with hefty promotional budgets.

The social world is defined by the ability of individuals to socially network with others, often reaching greater numbers of persons than many corporate campaigns. Witness the Dominos' employees, whose YouTube "vile viral video" spoiled a "50-year reputation" using just a camera and Internet access.[11] Millions of viewers watched employees at Dominos tamper with the food on the video they uploaded to You Tube. Domino's recovered, thanks in large part to the restaurant chain's quick response and savvy understanding of social media. They flooded the internet with counter-programming in a quick-reaction campaign lauded by public relation professionals:

> *"Domino's was the latest company to be on the wrong end of a "Twitter storm," a spontaneously formed digital mob that rapidly shares information. The company's swift response to the employees*

and its wider customer base, using the same Web sites and media that spread the video, has been praised by observers who nevertheless wonder if the company can emerge unscathed"(as cited in Capps, 2010).... When something blows up on the web, managing the PR fallout requires a fast, targeted approach. While you can't keep people from being stupid, being both fast and thoughtful in how you respond can make all the difference for those whose jobs are caught in the middle.[12]

A savvy social media organization has the opportunity to shape the discussion while tapping into a source of content that can generate valuable ideas (content again) for new products, changes to existing products, and growth. But this requires an organization to be alert, quick on its feet, and base its operations and marketing in the digital world.

Feedback is often difficult for us to handle. Someone asks your opinion, but do they really want your opinion or do they want reassurance? A frequent topic for comedy is the uncomfortable situation of being asked by a friend or loved one, "Do I look fat?" What do you do? The savvy social marketer looks beneath the surface, digging deep for answers and 'truth.' The questioner may be honestly asking for an opinion about weight, or may be seeking reassurance; context often provides insight. A social world is also a world filled with opinions, welcomed or not, and savvy organizations and individual employees have to be careful to react in a measured and deliberate way. As the Inside Indiana Business blog put it: "Everyone has an opinion, and many are worth listening to. Many are worth forgetting as well." It then goes on to advise: Don't just respond, instead offer "some kind of useful service or product to customers" during the course of the continuing conversation.[13]

WITH SIX COMPLAINTS, YOU GET EGGROLL IN YOUR EAR!

Savvy involves care. A savvy company social user is a careful communicator. Yes, everyone says they appreciate honest feedback but reality often intrudes when, say, the owner of a Chinese restaurant contacts and berates customers giving negative reviews. I had a team of students assist a Chinese restaurant owner in monitoring restaurant review sites as the first step toward managing the perceptions of diners while bringing the owner into the digital communications era. The students were horrified when the owner went on the sites and scolded customers for their reviews.

Feedback was also an issue when I led the team that resurrected the E-commerce website of a women's shoe company. We had dramatic sales increases, but also the opportunity for immediate feedback. This angered one of our designers:"How can they say those things about us?" "They're wrong, wrong, wrong about the design!" They just "don't know the meaning of fashion-forward." My answer: "They may not know what's fashion-forward and what's not according to our standards…but they just know what they like. And so, since we've made it easy for them to offer feedback, we need to show them they matter, that we care about their opinions and take them seriously." The expressions of care coming in response to feedback are also content, part of the Virtual Relationship Cloud hovering around an organization. Your handling of complaints and negative comments, of reviews and blogs that offer a 'truth' you may now want to face or you may not agree with, is critical in shaping content. Your responses quickly enter the digital universe and become part of the conversation. The public relations blog, The Flip Side, cites this example and offers appropriate advice:

If you take a few minutes to read the comments on a product review, you'll notice the comments turn into discussions among users. But how often do you see a company join the discussions? We noticed that QuickBooks Customer Care reps are engaged in product reviews on Amazon. com. In many cases, reps reached out to customers who left negative reviews by giving email addresses and asking the best way to contact the customers. Most Customer Care rep responses used these techniques:

- thanking the customers for their feedback
- writing responses in a compassionate tone
- acknowledging the customers' concerns or frustrations
- offering to work with them to understand the situation from the customers' points of view to find a solution

Instead of ignoring problems, QuickBooks worked quickly to try to fix them and address them directly with each customer.[14]

The lesson here is to be deliberate in response, as the thoughts and opinions expressed by individuals in your organization are quickly absorbed as content in the social network. Content is the stuff of conversations: Conversations capture the interaction and opinion trends among affiliates with respect to a company's products and services. Put yourself in a position to shape and define content relevant to your organization by enthusiastically and positively joining in the conversation. Approaches include crowd-sourcing, contests, standardized policies for customer service replies, personalization rules, etc. The savvy social user understands that content in a digital world is king. And queen. And emperor.

CONTENT IS A MEANS TO AN END

In the end, content is a way to achieve your corporate goals and add value to the organization's relationship with affiliates. You add value by pushing affiliates, from a mild relationship with you, to a passionate relationship, a hyper-relationship, that has them actively evangelizing on your behalf. The Peppers & Rogers Group refers to this as "cultivating customer advocates." [15] Keep in mind, in the end you want to create an intense affiliation, so intense that affiliates are eager to share their passion and involvement with your organization, its products, and services with others. The kind of content that surrounds you, which comprises a company's virtual cloud, shapes the perceptions of newcomers and pushes existing customers into more intense affiliation. Content (the combined weight of conversation and interaction) is the substance of the socially mediated material flying around an organization. This creates an image and subsequently shapes future conversation.

In content, everything counts. Tone matters, emotional approach matters, perception matters. Although most socially mediated conversation is casual, it is worth following the conventions of your business category for spelling, graphics, grammar, etc. It is worth checking your sources; statements of fact can be somewhat easily checked. And above all, stay positive; positive trumps negative when it comes to building online image. The Twittown blog offers this advice:

Everyone knows how important it is to try and come off as being as positive as possible when we deal with people in our lives. It is important that we try to give off a positive image to those we work with, our friends, people we go to church with, people we first meet. None of us want them walking away thinking that we're a complete jerk. We should carry this attitude over with us when we are involved with any type of social network online as well. Just because we're dealing with those who are in cyberspace and we can't see them, doesn't mean they don't visualize an image of us . . . The images that we build online are just as important if not more important than your offline image. Because once you have created an image online, it is out there forever. It is easier to rebuild your image offline if it has been tarnished than it is online.[16]

Let's summarize: Perceptions, attitudes, opinions, text, and images all become content in social media. All of this builds an image of an organization and the Virtual Relationship Cloud that swirls around it, affecting its reputation and ultimately, its effectiveness. Peppers and Rogers Group view this new digital interactive era as "a fantastic opportunity for companies" because it offers the chance to "encourage positive discussions."[17] They urge organizations to use content to touch their best affiliates through conversation, making them "customer advocates:"

Advocates work on a company's behalf to promote the brand, enhance its reputation, and in some cases, drive new business. They are also more valuable customers themselves, buying more and being less price-conscious than other customers. And most important, they connect with the company on a deep emotional level, which in itself is the best competitive differentiator there is.[18]

Advocacy results from the natural growth of a hyper-relationship through conversation. Each relevant and affective conversation intensifies the bond between an individual and an organization, the cumulative effect creates emotional ties that predispose the affiliate to react positively to offers and communications from the organization.

A MICKEY MOUSE WORLD

Individuals react to their environment in different ways, and content is a product of their response to the environment. You may change the way people see things incrementally and over time. Hence, the need for a two-pronged strategy (two mints in one, remember?) to shape perceptions. The virtual cloud surrounding an organization is much like a culture, and in the age of social media a corporate culture is not just internal; it extends to the external, to the characteristics of the types of customers it attracts and shaped by the content of their communication and conversations. One of the first events I was invited to attend when I left business and went into education was a dinner lecture, ostensibly designed to bring 'notable' academic thought to our little corner of the higher education universe. Much to my delight, the lecture was to be about the marketing of Disneyworld, catnip to a former marketing and operations executive involved for much of his adult life in consumer merchandising. Exciting, I thought, the magic of Disney marketers, an inside look at the customer service and promotional magic of the Magic Kingdom.

Unfortunately, I had failed to read beyond the promotional headline for the event and sat through an anti-capitalist, oh-isn't-it-a-shame-that-popular-culture-is-ruining-the-world shtick (Yiddish for "act," often applied to comedy). The entire lecture and the intermittent cheers were, well, Mickey Mouse. This academic, whose major achievements were talking Disney executives into allowing a behind-the-scenes look at the company's Orlando theme park and persuading an academic press to publish the resulting book (oh, you're going to slam those bourgeois pigs, here's a contract!), spent an hour explaining why Mickey, Goofy, and the gang were harbingers of the doom of civilization. I looked around the room, I realized Mickey and Goofy were just fine, thank you; rather, I was sitting with the harbingers of the doom of civilization, professors, many of whom were as joyless as the mystery meat on the plate in front of me posing as chicken.

Wandering around the room after the dinner and lecture, I realized that most of my conversations were with faculty who, like me, had pursued education as a second career. I became friends with a number of the instructors I met that night, our subsequent conversations usually taking place online, in emails, and texting. Social media became a way to converse with others with similar viewpoint and interests, like attracts like. Our conversations were shaped by our experiences, our affiliation with other professional faculty, and our shared view of the traditional educator, at least in our neck of the woods, as Goofy in cap and gown.

So what, you ask? Good question. Digital conversation follows all the rules of face-to-face in that some content is worthwhile for a certain type of person, while other content quickly traverses the boundaries and goes rocketing off into outer space. The Mickey-as-the-root-of-all-evil lecture bounced off my skull and went off in search of more hospitable heads in the room. The success of content depends upon the characteristics of the individual on the receiving end. Our nature, the nature of a group (the selected target segment(s)) and goals of the organization should then, shape the content of the conversation and intensity of involvement by affiliates.

You can increase the probability of successfully encouraging conversation by having broad policies for content generation with a specific type of affiliate in mind (e.g., always be positive, use contractions, whenever possible link our product to positive life experiences, etc.). The more homogenous your targeted segment is, the simpler it is to generate content guidelines. In addition, make sure that each content type (blogs, replies to frequently asked questions, ordering instructions, tweets, photo posts, etc.)"feels" emotionally similar to the type of individual being targeted. Consulting giant McKinsey & Company tells its clients that creating effective content requires a four-step social media process that can be summarized as follows:[19]

MONITOR. Routinely observe the content affiliates are generating. The McKinsey consulting group suggests routine monitoring of the buzz surrounding an organization is a core function of a savvy social company, citing the Gatorade sports drink establishment of "a 24 hour 'war room' tracking this information in real time, offering vital insights and shopper feedback."[20] This offers a peek into the real-time effects of the efforts of an organization. For example, you've sent out an appeal for donations to begin the year and the reaction is evenly divided between enthusiasm, "I've been thinking about you and the great work you do," and annoyance, "Give me a break, I'm still broke from Christmas and charity begins at home".

RESPOND. Researching the content and conversation, you discover your timing is off by asking for donations on the heels of the Christmas season, and start a series of emails that begin "I don't know what we were thinking…" Responding to content is a critical duty of the savvy social media user. McKinsey emphasizes the importance of responding to both " positive and negative comments," citing the case of "a fake photo on Twitter claiming to show McDonald's asking African American customers to pay an extra service charge, which the firm effectively dealt with via official statements and engaging influencers."[21]

AMPLIFY. But you don't just stop there, you make sure your market responses have a "social motivator" (an incentive to converse with others) that encourages sharing among your affiliates and with your organization. All corporate responses should trigger positive conversation that can be monitored and used to shape an organization's continuing marketing efforts. McKinsey& Company offers us examples:

> Gilt Groupe, the online fashion site, and Groupon, the group buying platform, achieve this by giving shoppers credit when they refer a first-time customer. McKinsey found direct recommendations deliver engagement rates 30 times higher than digital ads. Starbucks also showed how to foster loyalty after purchase, having launched a contest promising a $20 gift card to users uploading pictures of its new ad billboards on Twitter. Equally, customer service communities have become popular, with Intuit, the software specialist, creating forums for its Quicken and QuickBooks products, where consumers answer 80% of questions.[22]

And so (back to our charity scenario) you begin, say, a new donation drive in which gift givers send donations of each friend or family member in whose name they are giving a donation for a new Our Friends & Family microsite you've created for this campaign.

LEAD. Now that you understand the content of the conversations, how and why your affiliates are reacting the way they do, it is time to get in front of your market and "encourage behavior change."[23] Leadership is as important to content generation and management as it is to running an organization. Another example covered in the McKinsey ^& Company study discussed above:

> Old Spice's "Smell Like a Man, Man" campaign, which yielded 19m hits and saw sales rise by 27% in six months. Bonobos, an online menswear company, logged a similar result through providing discounts to its Twitter fans that retweeted its messages a certain amount of times, bringing in 100 first-time buyers and securing a payback of 1,200% in one day.[24]

I CAN'T GET NO SATISFACTION!

Keep in mind that a goal is to create connected consumers passionate about their relationship with the organization and each other. This passion is displayed in the content of their communications, continuing nature, and emotional depth of their conversations. While satisfied customers are desirable, we need to go far beyond satisfaction to engagement, involvement, and psychological investment. We need to lead an affiliate to loyalty, to the organization, and its products and services. Satisfaction is fleeting, often lacking emotional investment. Take a look at what the Rolling Stones had to say about their feelings for a company that failed to bind with its products, to develop a connection through continuing conversation and interactivity. The organization did not offer a path to hyper-relationship, relying on traditional (or, as I like to put it, Old Media) media to have an impact:

> When I'm drivin' in my car
> and a man comes on the radio
> he's tellin' me more and more
> about some useless information
> supposed to fire my imagination.
> I can't get no, oh no no no.
> Hey hey hey, that's what I say.[25]

A radio ad? A 15-second or 30-second spot about a product? This song segment illustrates the futility of traditional advertising, of passionless marketing, and content floating free of an affiliate anchor. No wonder Mick Jagger's face has more lines in it than a Walmart on Black Friday! He can't feel the love! And he certainly can't feel the satisfaction.

The savvy social organization understands that mere satisfaction is not enough in the age of social networks. Peppers and Rogers note:

> Improving customer satisfaction is no longer enough for companies looking to develop meaningful relationships with customers. Satisfaction is the base expectation of a customer. Therefore, it doesn't stand out in the customer's mind. Today real customer strategy is about creating customer advocates. But can advocacy, which revolves around emotion, be operationalized? Yes. It's about formalizing the drivers of an emotional connection. Customer advocacy is a lofty goal, but can be achieved as part of a multilevel customer experience strategy using three customer experience triage tiers: Silence the detractors, build a solid customer experience, and develop advocates.[26]

Content should produce more than satisfaction, it should bind. You must understand not only how your affiliates communicate, but also the content approaches that best fit with their lifestyles. Some rules for content:

1. USE LIFESTYLE-FRIENDLY SUBJECT MATTER AND STYLE. DailyCandy, for example, entered into a marketing partnership with NBC Universal's Style Media cable network solely on the strength of the trendy women's site's ability to relate to its young, urban, and professional affiliates. The publicity release spelled out the added value that the website's editorial content is bringing to the table:

> From the inside scoop on hot new trends to expert advice on how to throw the ultimate dinner party, StyleCandy will deliver everything women need to know about the latest in fashion, beauty, lifestyle, technology, and pop culture via short-form video delivered across Style platforms (both linear and digital) and via DailyCandy emails and DailyCandy.com. Together with Style's on-air talent, DailyCandy Chief Correspondent SuChin Pak will be among the contributors for StyleCandy. "This is an exciting opportunity to bring together two of the most powerful brands, which have highly complementary and passionate communities, in the women's lifestyle space," said Nick Lehman, President of Digital for NBCUniversal's Entertainment & Digital Networks and Integrated Media. "Today's audiences are multiscreen consumers, and StyleCandy creates a seamless experience across TV and digital -- something NBCUniversal is uniquely positioned to provide." DailyCandy and Style Media's large, combined social media following of more than 4.8 million will also engage with StyleCandy's original content through Facebook, Twitter, Pinterest, and other key social media platforms. Future phases of the collaboration will deliver optimized content to mobile users.[27]

The content speaks to the consumer through tone, style, and subject matter across media categories. It is digital, interactive, and trans-media (same narrative customized to different media) in approach.

2. USE A PERSONAL TONE. Customers are individuals. As much as an organization thinks of its target as a group, it is, in fact, a group of individuals, each of whom is attracted to the security and warmth of a personal relationship. There are few things as off-putting as having an emotional connection and then getting hit with stiff, formal language that reads like a thesis or term paper. Leave the stiff stuff to the professors, personalize your online presence throughout your hub-and-spoke. Address your affiliates directly and with cheer. Anyone can do it in a social world. It just means "letting your hair down," as this manager in an accounting trade organization advises his accountant-world members:

> In the ever-changing world of online marketing and public relations, businesses are now utilizing social media to interact with consumers in a very personal manner. As a result, long-lasting bonds are being created, and these connections are increasing customer loyalty and satisfaction. Social media is a great tool for businesses to use to cultivate customer relationships and learn how to engage their audiences, but businesses also have a lot to consider when using social media. Below are several tips explaining some of the best practices for companies using social media to connect with their clients.[28]

Even accountants are using personalized content in their social marketing, is this a great country or what?!

3. ACTIONS SPEAK AS LOUDLY AS WORDS AND YOU NEED BOTH. Here we draw on the wisdom of direct marketing, where offers are, like social content, king and queen and emperor. Talk, yes, but also talk *about* something to do and make an offer. Your strategy should include a steady stream of offers in your social content: Buy now…, order now…, tweet this to your friends…, reply to this email…, "like" us on Facebook and… you get the picture. Consider how the following top eight direct marketing offers collected by Dean Rieck of MelissaData can be adapted to your social strategy:[29]

> 1. Free Trial. This may be the best offer ever devised.
>
> 2. Money-Back Guarantee.
>
> 3. Free Gift. When you offer a freebie your customer wants, your offer will usually out-pull a discount offer of similar value.
>
> 4. Limited Time. An offer with a time limit gets more response than an offer without one.
>
> 5. Yes/No. You ask your prospect to respond positively or negatively.
>
> 6. Negative Option. This pulls better than positive option offers. You offer a free trial or a special deal on a product then automatically ship future merchandise unless the customer specifically takes an action to refuse.
>
> 7. Credit Card Payment. Nothing is easier than paying with plastic.
>
> 8. Sweepstakes.

4. SPEAK ME SOME CHRISTIAN, OR JEWISH, OR ACCOUNTANT, OR TRENDY.
When your affiliates are Christian, speak Christian. Use the word "blessing" once in a while. If they're Baptist, talk casseroles and black bean salad; if they're Methodist, talk socialism; every denomination has its unique culture and method of expression. When you're developing membership programs for a Jewish Community Center, break out the bagels, lox, and New York Times. Accountants? Try dull, it works! New Yorkers? Be blunt and to the point.

5. PRODUCTS ARE PEOPLE, TOO. Personalize your product and service images, text, and presentation. Product descriptions should be short and to-the-point, focusing on the benefits for the affiliate. Customers are customers because they *buy* something; donors are donors because they *donate* something; students are students because they *register* (and pay) for courses. The more attractive the content relating to the object and/or activity of the transaction, the more likely the affiliate is to feel connected. Think of your products and services as animate, full of life, and ready for relationship with your affiliate. A pair of shoes on your E-commerce site should be excited to be on the feet of your customer, and should show it in the content, the combination of graphics and text. Mission donations are not monetary transactions; they're smiles on the faces of children, hopeful feelings, and service to something greater than oneself. Personalization works! It fits the way we, as humans, interact with the world. We take people, events, and things *personally*. Have you met my car, Arthur? Let me introduce you to my vegetable garden, starting with Edgar the Eggplant. Perhaps it's crazy, but also human.

5. TESTIMONIALS R' US. Some of the most effective content is testimonials, the narratives of real people who have used a product or service, joined a church or synagogue, visited a tourist attraction, and then *enthused* about it. People are naturally attracted to the experiences of others as they contemplate following in their footsteps. That's why the tremendously successful grocery delivery service showcases the love their customers have for the online ordering service. It's not just a grocery service; it's social, a topic of conversation, a community of true believers in the company and its products, and enthusiastic advocates:[30]

> *"Not only would I recommend you to friends, but I have talked you people up to virtually everyone I know in New York. You guys are the best invention since ice cream!"* -**Rosina, Upper East Side**

> *"You blow the (non-existent) competition out of the East River!"* -**Adrienne, Ridgewood**

> *"Thanks for being a terrific service that has allowed me to eat better, cook better and spend less time running out for groceries and household products!"* - **Teresa, Forest Hills**

> *"My orders are on time, and the food is of excellent quality. I would be miserable if I had to move to a neighborhood that did not have FreshDirect!"* - **Kristiana, Astoria**

THE MARKETING HANDBOOK BLOG SUMMARIZES THE COMMUNITY APPEAL OF THE SERVICE AS REFLECTED THROUGH ITS ONLINE HUB:

> This isn't your usual online grocery retailer. The site romances food with luscious photos and lush descriptions, making the pages a virtual feast for food lovers. Visitors can take a photo tour of the facility's food departments and browse mouth-watering photos of produce and premade entrees. FreshDirect also posts customer testimonials, a nice bit of word-of-mouth, and offers recipes too. Not surprisingly, FreshDirect enjoys unusually high customer loyalty. Its ads and celebrity shopping lists have a distinctly New York flavor. Who outside of NYC still remembers former mayor Ed Koch? Warning: Don't look at this site when you're hungry.[31]

THE TAKE-AWAY

Content counts, a whole bunch! The savvy social media strategist is deliberate about shaping the content generated by both the organization and its affiliates. All content matters, and all content provides valuable feedback that allows a company to keep its finger on the pulse of its community.

We offered some rules for content. Most involved personalization: We're human and humans personalize the world. We assess our activity and conversation based on how it makes us feel *personally*. Content is generated by both organizations and affiliates. The savvy social organization monitors the content, pulls from it insights for operating and marketing, and shapes the content according to its goals and the benefits sought by affiliates. We celebrate and, in fact, we advocate the products and services that gather a celebratory and affectionate community, like Starbucks, your local church, or even a favorite brand of car. Whether hub or spoke, blog or microsite, content is what draws people to your site. People buy, people donate, people talk, converse, and join. Content is generated by people for people, and the more personal, the better.

Content is, indeed, king, queen, emperor; and, for that matter, czar, czarina, and czardine. Heh.

(ENDNOTES)

1   http://blog.nielsen.com/nielsenwire/online_mobile/social-media-accounts-for-22-percent-of-time-online/?utm_source=feedburner&utm_medium=feed&utm_campaign=Feed%3A+NielsenWireMediaEntertainment+%28Nielsen+Wire+%C2%BB+Media+%26+Entertainment%29&utm_content=Google+Reader

2   http://thesocialskinny.com/100-social-media-mobile-and-internet-statistics-for-2012/

3   http://royal.pingdom.com/2012/01/17/internet-2011-in-numbers/

4   http://www.yelp.com/biz/mesa-grill-new-york, accessed June 6, 2012

5   http://www.1to1media.com/view.aspx?docid=33497&utm_source=1to1weekly&utm_medium=H&utm_campaign=03262012

6   http://www.cbsnews.com/8301-505125_162-51052961/what-is-crowdsourcing/

7   http://www.forbcs.com/sites/melaniewells/2010/11/18/harley-davidson-cmo-crowdsourcing-ads-to-engage-a-new-audience/

8   http://www.powersportsmarketing.com/nocages/

9   http://www.crowdsourcing.org/document/why-crowdsourcing-can-work-for-harley-davidson/8366

10   http://www.1to1media.com/view.aspx?docid=33543

11   http://abcnews.go.com/WNT/video?id=7358663

12   http://earlcapps.blogspot.com/2010/03/managing-viral-pr-damage-dominos-pizza.html

13   http://www.insideindianabusiness.com/contributors.asp?id=2141

14   http://www.theflipsidecommunications.com/2012/02/public-relations-handling-negative-customer-reviews/

15   Matthew Roden, "Cultivating Customer Advocates: More Than Satisfaction and Loyalty," White Paper, 2011, Peppers & Rogers Group, Stamford CT

16   http://twittown.com/social-networks/social-networks-blog/importance-creating-positive-online-image

17   Roden, p. 2

18   ibid

19   http://www.warc.com/LatestNews/News/EmailNews.news?ID=29750&Origin=WARCNewsEmail

20   http://www.warc.com/LatestNews/News/EmailNews.news?ID=29750&Origin=WARCNewsEmail

21   Ibid

22   Ibid

23   Ibid

24   Ibid

25   http://www.lyricsfreak.com/r/rolling+stones/satisfaction_10243634.html

26   Matthew Roden, "Cultivating Customer Advocates: More Than Satisfaction and Loyalty," White Paper, 2011, Peppers & Rogers Group, Stamford CT

27   http://thefutoncritic.com/news/2012/05/01/nbcuniversals-style-media-and-dailycandy-two-of-the-most-powerful-brands-in-the-womens-lifestyle-space-forge-multiplatform-partnership-389000/20120501style01/#w EJevxp0X7EqHxJF.99

28   http://www.nasba.org/features/businesses-getting-personal-through-social-media/

29   http://www.melissadata.com/enews/marketingadvisor/articles/0607c/5.htm

30   http://www.freshdirect.com/about/testimonial.jsp?catId=about_test

31   http://marketinghandbook.blogspot.com/2009/02/fresh-direct-makes-mouths-water.html

CHAPTER 5

TACTICS:
HOW MANY WAYS CAN YOU SAY, "I LOVE YOU?"

INTRODUCTION

LOVE. REAL LOVE. ORGANIZATIONAL LOVE. DIGITAL LOVE. CORPORATE LOVE. COURTNEY LOVE (OOPS, SCRATCH THAT, I GOT CARRIED AWAY).

In the jargon of marketing communications, interactive strategy is about "building affinity." But in plain language, it is simply saying, "I love you" to a customer over and over and in as many ways possible. Love is the foundation of a continuing strategy that encourages a coming together of an organization and its affiliates, and shows in the care taken by the organization as it engages the customer. From beginning to end, the savvy social marketer uses a multitude of ways to let an affiliate know that he or she is not alone in a relationship with the organization, that it cares, and that everything in the relationship is designed to increase the depth of connection. The organization is concerned with a customer's future, got it?

Words, graphics, and actions together invest the relationship of the organization and individual consumer with meaning beyond the simple act of buying, donating, or shopping. A savvy social and growth-oriented company makes sure that each moment of truth (contact of an individual with the organization) does more than promote positive feelings. Rather, each touch should convey the desire by the company for each affiliate's well-being and growth as an individual. C.S. Lewis, author in the pre-digital era (who created the fantasy world of Narnia), put it best: "Love is not affectionate feeling, but a steady wish for the loved person's ultimate good as far as it can be obtained." [1] In the digitally-rendered 21st century, we have the tools to spread genuine love in service of corporate growth, a strategy that, more often than not, contributes to out-of-the-ordinary growth. Don't believe it? Forbes annual list of the world's most admired companies reads like a *Who's Who* of corporate growth pacesetters. [2] Around the world, individuals have told Forbes that these are companies that have caused them to feel wonder and confidence in the organizations' wish for their well-being, causing them to regard these organizations with pleasure and approval. These companies care for their affiliates and let them know it: the overwhelming majority are savvy and imaginative users of interactive digital media; collectively, their use of social media to advance the qualities that make them most worthy of admiration by customers and affiliates around the world also reads like a strategic primer on business caring, sharing, and concern for customers.

Everything the savvy social media user does in the marketplace is designed to let targeted segments know that the company cares—really cares—about their lives, present and future. Inbound communications through the spokes are greeted with an empathetic "let me make your after-our-conversation life better than the before-our-conversation life." Offers are designed to connect with the preferences and needs of customers. Editorial content is created in easily engaged modules, allowing the company to display an understanding of the time demands on the life of affiliates in the digital era; and interactivity is based on the behavior and values that come about with a "love" of social media strategy. We care, and our conversations, our offers, and interactions are an extension of that care, our love. A growing and savvy organization understands that behind every touch with an affiliate is an individual in the organization who understands his or her responsibility to demonstrate an interest in the life of the affiliate with whom he or she is dealing. While it is easy to love a class of persons or things ("I just love ice cream!"), it is not always easy to love individuals or specific products, services, and experiences ("Oh, that bubblegum ice cream from Ye Awesome Creamery Shoppe is detestable," proclaims I-love-ice cream guy). An organization is corporate, but touches with affiliates are individual. In the end, the relationship is the sum of individually shaped moments when an organization bares its soul through a single person, an individual interacting—touching—a customer. That is why exposures of an organization to individuals are called "moments of truth." Businessdictionary.com defines "moment of truth" as an "instance of contact or interaction between a customer and a firm (through a product, sales force, or visit) that gives the customer an opportunity to form (or change) an impression about the firm."[3] Savvy and growing organizations show respect and care for the individuals with whom they come in contact. Love becomes a deliberate strategy: Japanese automaker Subaru, for example, cruised through the latest economic downturn largely on a customer focus that emphasized "share the love" and a corporate culture that emphasizes the creation of "love between the owner and the automobile" and the organization it represents.[4] Love, in a socially mediated world, needs to be part of every strategy, pushed down and carried out by individuals trained to show care for others through digital conversation and tools.

An organization must deliberately 'push down' the importance of love to its employees as a basis for dealing with customers in every portion of corporate life. Growth requires a critical mass of employees who share affection and concern for customers of their company, and the ability to put the occasional (or sometimes, it seems, frequent) intemperate customer in a larger perspective. Examples:

- Sure, we failed to live up to your expectations and we're working to remedy that, give us a chance!
- I will patiently listen as you scream in my face (and will even call the paramedics when you give yourself a heart attack!)
- What do you mean, "the customer is always right?" Not when you want to return a toaster oven you bought five years ago from a competitor.

Love is not easy, especially when you represent an organization. Every expression of love throughout history sooner or later demands qualities that take effort and discipline to summon from the human soul. To do so, an organization must emphasize training, mentoring its employees in the ways and means of getting across to affiliates the notion that the organization—and all the individuals that comprise it—uniformly cares about the future of the individuals associated with it.

Again, it comes down to the "touchy-feely" stuff that training and leadership can get across. Most people are not naturally touchy-feely (oriented toward reaching out and caring about the future of others) and it takes an intense and continuing effort to keep us on the touchy-feely track when so many moments of truth involve difficult and exasperating individuals.

THE LOVE BOAT

Yes, *The Love Boat* was a classic seventies situation comedy on television that featured the romantic escapades of the crew of a cruise ship. The plots varied over its ten-year run, but the underlying theme remained the same: it is possible to be a loving person, but you're going to have to work at it, put up with a lot of goofy characters, and you'd better get it done quickly, because we only have forty-two minutes when the commercials are removed from an hour show to summon our inner *Love Boat* and make an impression before the show ends!

One of the most difficult parts of savvy social media is achieving a loving approach by employees of an organization. The display of deep-seated concern and affection for another demands a continuing effort to work through the continuous conflicting emotions that intrude on our relationships with others: Annoyance, anger, disappointment, boredom, resentment, moodiness, you get the picture. When you are displaying love on behalf of an organization, well, it gets complicated. I may feel loving, that all's right with the world, but this customer in front of me (or online, as face-to-face and online communications are all part of the corporate VRC, or Virtual Relationship Cloud) was just ignored by my associate who thought lunch was much more important and now I'm left to deal with the mess. Love, ri-i-i-ight. This is gonna hurt!

So you take a deep breath and summon every reserve of strength and discipline that you have and smile at a customer who has been ignored, who expected better treatment, and who may have to take indifference from her children, from an employer who looks right through her, and a husband whose eyeball have been glued to ESPN for a decade, but she'll be gosh-darned if she'll take it from this store or website helpline. Notice, it's the store, the company, the organization; when push comes to shove and affiliates are affixing blame, even though it was an individual who treated her poorly. Organizations are represented by individuals, and it is individuals who do the sowing but organizations that do the reaping, suffering the consequences of an individual's lack of love.

Love is work, love is often agonizing, and love demands sacrifice by one individual for another. Yes, it has its rewards, but much of love is difficult. Take a look at what at least one entertainment website, kidzworld,has called the top romantic movie of the past decade: *Twilight*. The description of Bella and Edward's loving relationship is not exactly the stuff of fluttering birds and singing dwarves:

> *Life sucks when your soul mate is a hundred-year-old vampire who has to resist the urge to literally eat you up. Considering the age gap and diet differences, Edward and Bella manage to do alright with the whole doomed lovers deal—sure there's pining and wistful sighs, but those dramatics are part of the reason Twilight has staked the number one spot in so many hearts. 5*

But that's my point exactly; love in the context of the social media marketing of organizations is exactly what love is in the lives of individuals: difficult, tough, trying, challenging, demanding and at least a dozen other synonyms that a quick Google search would yield. In fact, Google "love hurts" and you get 49.2 million hits; Google "love thrills" and you come up with 25.9 million. Sure, the love of an organization for its affiliates, as demonstrated by the millions of contacts each day between individuals representing the organization and its customers, demonstrably adds value and growth to a company. But our corporate soul mates (customers) often resemble blood-sucking vampires. Love *that*, why don't ya!

So what's a savvy social user to do? Strengthen your spine, realize that dysfunction happens, both in affiliates and organization, and take some advice from HubSpot, which emphasizes the benefits that come from a deliberate strategy of lacing all of your social media efforts with love:

When a business regularly shows its customers it loves them, that business also becomes a lovable brand.

And making your brand lovable has a ton of great business benefits, including:

- Customer Referrals
- Customer Evangelism
- Social Sharing
- Press-Worthy Testimonials
- Customer Retention and Loyalty

Think about a brand you love. Let's call that brand 'Company X.' Chances are, you'd tell a friend to try that brand's products/services (customer referrals), you'd talk about how much you love them when the topic arises (customer evangelism), you're happy to share interesting content that brand creates with your social networks (social sharing), you'd be pleased to have that brand feature one of your "I love Company X" rants (testimonials), and you regularly choose this brand over its competitors (customer loyalty). [6]

What does this mean? It's not easy, that it takes work and self-discipline, and that, for an organization, it is worth the investment in time and resources to train the individuals within a company in loving its affiliates, especially those handling social media.

Love adds value to a savvy social organization because it brings results, presaging more intense affiliation and loyalty, and a customer base willing to overlook the inevitable lapses and dysfunctions of every organization. Tim Allen, star of the half-hour comedy-with-a-heart television hit, *Home Improvement*, noticed something about love: "A guy knows he's in love when he loses interest in his car for a couple of days." In other words, love has a way of involving an individual in such a way that other parts of life (issues, opportunities, events, and pastimes) fade into the background, leaving us free to concentrate on the relationship, or, more specifically, the hyper-relationship.

For the savvy social organization, the touchy-feely stuff ends up as an investment that promises greater returns. A company's representatives who reflect love, as difficult and imperfect as that may be, help affiliates focus on their relationship with the organization, encouraging them to do more, buy more, donate more, increasing their activity with and on behalf of the organization. A deliberate strategic approach that incorporates policies for folding love into content and technology adds value to the relationship and brings a greater probability of growth in a social world.

Yes, it's hard. Who's to say that a graphic on a website, a software design feature, a particular application, a live chat script displays love? That a returns policy demonstrates love? That a post, tweet, or photo has an underlying love for a customer? Of course, traditional marketers, told to incorporate love into their every activity, will look at you as if you're John Carter and you've just returned from Mars. Love, really!? Going back a chapter or two, "Aw, c'mon man!"

TO TRAIN OR NOT TO TRAIN

William Shakespeare penned one of the more famous monologues in literature through the mouth of Hamlet in "The Tragedy of Hamlet, Prince of Denmark." Hamlet was a gloomy and unstable sort who walked around Denmark in tights and sulked his way from one tragedy to the next, much of it avoidable if he accepted the wise advice of others (which he did not, being Danish royalty and all). Everything seemed to go wrong with him; not unexpected for a grown man partial to tights in a country where winter lasts eleven months. One day he threw up his hands and gave us an English literature professor's version of "I've had it!"

To be, or not to be—that is the question:
Whether 'tis nobler in the mind to suffer
The slings and arrows of outrageous fortune
Or to take arms against a sea of troubles
And by opposing end them...

"To be or not to be? That is the question." What is the lesson in this? Yes, don't wear tights in sub-zero temperature, especially if you're a man. More to the point, people, if left to their own devices, will go their own way and quite often make a hash of things, even in the simplest and most obvious of circumstances. The key here is "left to their own devices." Training, education, mentoring, all the tools of successful companies, are also a critical part of successful social media marketing. Members of an organization, from the janitorial staff to the senior executive team, need training and advice from others. Training and operational policies help to sharpen the focus of individuals within an organization, constantly reminding them of the need to concentrate on a certain set of goals. That is just how we are: Humanity, all of it, needs guidance.

Without guidance we go with instincts that are often inadequate to the task at hand. I once hired a customer service manager whose tenure was disastrous. I put her into the role without training, assuming she understood our goals and approaches, huge mistake on my part. More than g 90 percent of our customer service calls to the 800 number listed on our website went unanswered during her (short) time at the company.

"I don't understand this," I said, paying a visit to our customer services offices one night. "Your employees are all out in the parking lot having a grand old time—and all I hear is silence in here." I surveyed the empty desks where dozens of employees were to (sadly, theoretically) serve and reassure the young women that were fanatical buyers of our trendy shoes that we care.

"I like to give them a break," she said, "It keeps up morale. They get up every half hour. It's hard to keep alert in such a demanding job without a break."

"But who answers the phones? The phones should be staffed every day, every minute."

"I just shut down the phones when we're away."

"You shut down the phones?!" I repeated incredulously. "A customer service manager doesn't shut down phones. A customer service manager solves problems for customers, helps them, lets them know we care and love to serve them, keeps our employees focused on customers."

She looked at me. "Well, I didn't know. No one ever told me."

And I had not. I just assumed she understood, from the job description and title, that creating happy customers would be the focus of her leadership efforts. The first part of creating a loving response is training all those who work with customers, reinforcing the need for all 'touches' with affiliates need to display love. Digital words, online and offline graphics, service manners and mannerisms all should reflect love. Love is the preferred response. In an earlier chapter, we discussed the notion that all individuals want to feel loved, to feel that he or she matters. In social media, digital strategies are based in large part on designing love into every part of the communication chain so that, in a personal and digitally mediated way, all individuals affiliated with the company feel that they matter.

The savvy social media strategist understands that the digital era provides the tools and bandwidth (figuratively and literally) to personally and intimately addresses all of the affiliates of an organization. You expect nothing less as a customer, sensing, if not knowing, that companies have the ability to instantly collect enough information about you so that you're not a stranger, and thereby be able acknowledge your individualism in their dealings with you. The expectation is that an organization will, in some way, shape, or form express interest in you as an individual and demonstrate concern about your future.

How does a savvy social media manager put that into operation? How do you satisfy an individual affiliate who is asking to be treated as an individual who matters? Begin by paying attention to the individual *representing* the target, not as *representative* of the target. The former is a person with likes and dislikes, history and future, frustrations and ambitions, an individual; the latter is a class, a marketing target, a bloodless category. Get personal, understanding that individuals represent the organization to other individuals with similar hopes and dreams. In our personal lives, the more we grow in attachment to others, the more likely we are to demonstrate and express continuing concern for their well-being, and they for ours. Think of socially mediated relationships as personal relationships writ large, organizations as a collective individual whose words, actions, and attitudes must touch individual affiliates if they are to achieve the intimacy of hyper-relationship. Design entrepreneur Michael Hess, who has successfully built a business using talent, discipline, and digital interactive marketing, put it this way; "No matter what procedures, processes, people or tools you put in place, empathy—the ability to identify with and understand somebody else's feelings or difficulties—is a quality without which superior customer service simply can't exist."[7]

He then goes on to lay out questions to ask about intimate and effective customer service, a corporate love for individuals that can propel organizational growth and, thereby, opportunity for both the organization and the affiliate. All of the questions underline the need for employees, acting on behalf of an organization, to have the tools and attitudes that will allow them to love the customer:

- How does the person I'm trying to help feel?

- How would I feel if I were that person?

- No matter the request or the "rules," is there something I can/should do to help?

- What would I expect to be done for me if the roles were reversed?

- In the end, what would make this customer satisfied or (better yet) happy, and is there any reason I can't do it or find someone who can?[8]

In real life—your life—putting yourself in another person's shoes *works*. Trite but true, right? My wife snaps at me. I'm ready to return a few sharp words and suddenly I realize: She's weary, just back from two weeks of caring for an extraordinarily active daughter recovering from hip surgery. Slow down, think, project yourself into the other person's life right now, and you find yourself less likely to return fire. It works the same way in an organization, but with the added advantage that a savvy growing organization has education and policies in place for training employees in loving social tactics. Say this in this circumstance, that in that circumstance, smile and the world smiles with you; the stuff of training and mentoring.

FAQ 11: So let me get this straight: You're including training in love—LOVE!— as part of the responsibilities of a savvy social strategist? Isn't this a bit...weird?

Short answer: Yes. Long answer: Yes. Sure, I'm adding to this savvy social stuff, the notion of training your staff in generating content and service that reflects the love of an organization for its affiliates. Yes, this is a bit weird by conventional corporate training standards, but conventional training standards are so...yesterday. Yesterday didn't have everyone looking at you, watching you, assessing you. Yesterday you went out, found a job, did it—sometimes well, sometimes not so well—and went home. You put your time in. But this is now and you, dear Reader, need a competitive edge in a digital marketplace that is virtually everywhere. You have value in this vastly more competitive digital world, to the extent that, you can assist in growing your company, whether non-profit or business. Part of adding value to you as a candidate for a job, helping you to stand out, is providing a lesson in the softer side of savvy social media use. Companies often ignore the needs of customers for caring and sharing; you're an asset when you bring an understanding of the strategic value of customer love. You'll stand out, as tradition, for one thing, dictates against a touchy-feely approach to customers.

Corporate goals tend to center around *counting stuff* rather than emphasizing the social quality of the stuff being counted. Count the number of customer service contacts during the course of an hour, for example, rather than measure the number of customers who left the phone or live chat feeling loved and included. Consequently, organizations often neglect to train employees on attitude. Some approach customers in a loving way, most do not.

However, you are reading this blext and, therefore, understand that love is as much a part of corporate life as a part of personal life. Think about it: We spend much of our time looking for satisfying relationships, why should the organization/affiliate relationship be any different? Love has a direct line to growth. Direct marketers discovered this a long time ago, using the RFM approach: Recency, Frequency, Monetary. They measured the value of a customer and the strength of a company's relationship with its best customer on the basis of how recently (R) the customer had bought (or donated), the frequency (F) of the purchase or activity, and the amount (M) spent. Easy to draw a conclusion: The more a customer spends, the more recently he or she has visited and purchased, and the more times he or she interacts with the organization, that's how much he or she feels loved. Money can't buy love, but people can when they're trained!

Growth-oriented companies emphasize training in the implementation of both empathy and results through social media, a must for success in the digital era. The training and policies of Hess's company, Skooba Design, emphasize compassion and results. The hub of his online effort, the company website, is deliberate in letting customers and prospective customers know, through text and graphics, the *love* the company applies to customer relationships. From the About Us section, on Skooba Design's website,

> *[O]ur profession—and our passion—is designing exquisite cases and accessories for tech-travelers. We strive to marry quality and style with the necessary protection of an equipment case, and to create new and innovative features that you won't find anywhere else. Features that simply "make sense" when you see and use them. We have over 15 years' experience in designing and making protective cases and bags, an in-house product development department, and a real love of what we do (and who we do it for). And although we are not in it for awards, we have been honored with quite a few.9*

The organization has "a real love of what we do (and who we do it for)." Touchy-feely: touching the customer and provoking feelings of warmth. The recurring theme? You matter. We care about you. We love you. And BTW, although rewards are secondary to our relationship with you, we've won quite a few and that makes us even better able to give you more, share more, and love you more. The company uses content to emphasize individual relationship, addressing customers and prospects in the first person, using the word love, and tailoring both backend and frontend online functions to show that love.

SAY WHAT!?

What you say makes a difference in a digital world, as does how you think. The social media era is the age of thought and attitude instantly expressed, of a myriad of nonverbal and attitudinal cues no longer hidden but laid bare by instant and rich communication. How you are represented in text and graphics; the applications you choose to use, the convenience for the affiliate built into your software; your site navigational tools; the benefits waiting at the end of interactivity. All of these and more make a difference in customer response, both now and going forward. Before anything goes online, it is appropriate to ask if, in any way, what you are putting up is in the least bit denigrating of your target segments. In other words, question everything with a firm, "Say what?!"

The social media world is ablaze with putdowns, sarcasm, and snarks (snide and remark combined into a single adjective). Journalists, especially, have fostered a culture of cleverly phrased destructive remarks. Again, this is no different than relationships in our personal lives, where we tend to avoid people who are sarcastic and cynical while embracing those who are genuine, caring, and display interest in others and us. With them, it is not "all about me" all the time; it is occasionally and, perhaps, largely about others. When it is about others, relationships grow. When it is about me, the relationship doesn't go anywhere.

We are interested in growth, connection, and relationship. So forget the snark, the smart remarks, that hold few, if any, benefits beyond a single moment that shows how clever you are but leaves affiliates angry and empty. You want to grow your organization, not your reputation for laughter at the expense of someone else. Many individuals use social media to poke at others. Social media marketing results, at the corporate level, depend upon a deliberate strategy of connecting through the use of love and tactically expressing that emotion throughout each *hub* and *spoke*. Some put-downs that have appeared on Twitter, for example, illustrate a worldview that has no place in successful digital marketing or any kind of lasting relationship:

- If loving you is wrong, I've probably loved you
- Casual Friday is a great way to confirm your suspicions on who the sluts are at work

- Every time you watch Jersey Shore, a book commits suicide

An organization that is pursuing growth should, through training and leadership, make sure that content creation solves problems and, based on love, displays an essential respect for the individuals involved with the organization. It is easy to see exactly the type of person you do not want, go to YouTube or Tumblr for examples, and watch videos and read comments retail employees put up about customers:

> **Client:** I can't open your document…
>
> **Me:** Do you have Adobe Reader? You need it to read PDF documents.
>
> **Client:** PDF…? How do you spell that?[10]

The savvy social marketer asks: Is that the kind of person I want working for me? Someone who will play a significant role in undermining the growth of my organization, my future opportunities, and compensation? Dale Carnegie, put it best: "Any fool can criticize, condemn, and complain—and most fools do."

The first step in getting effective content out of your marketing communications staff is establishing love as part of the mission. Somewhere, somehow, an organization emphasizes that affiliates will leave each moment of truth feeling better about the company and better about themselves. It doesn't mean that everyone has to do the touch-feely stuff all the time. It means that the individuals who produce social media content on your behalf put themselves in the place of the customer, and leave feeling good about themselves and the customer.

JUMPING THE SNARK

The fifth season of the 1970's television hit *Happy Days* featured the popular and cool character, Fonzi, jumping over a shark tank during a visit to Hollywood. Some critics maintain that this gimmicky episode, which has since become part of the culture when someone or something is no longer relevant and current, began the downhill slide of the show. It had "jumped the shark." In effective social media for an organization, snark has jumped the shark.

Sarcasm, arrogance, anger, the digital landscape mirrors the human landscape. Snark is ubiquitous but tiring to the average person. We want connection, empathy, caring, not anger 24/7. How about another trite but true statement, this time from the realm of philosophy: There are no bad social media, just bad people. Social media as a function is neutral, but in its form (features of digital media such as ease of use and ability to throw relatively unfettered thoughts and impulses out there) it is personal in nature and immediate in action. In the days of Old Media a person would, say, write a letter. This took thought, consideration, and coordination because a letter meant physically acquiring an envelope, a stamp, pen and paper, finding an address, and taking a trip to a post office or mailbox. Each stop on this journey of expression took time and distanced the individual from the event (good or bad) that created this urge to write.

Our natural instincts can now be expressed at the click of a mouse or tap on a smartphone, it couldn't be easier. As a result, we can quickly do the wrong thing, the rude thing, snap angrily. Savvy social media use also sets up a process to control our baser instincts and build on our urge to individual meaning. The great classical scholar and Christian writer C.S. Lewis said, "Affection is responsible for nine-tenths of whatever solid and durable happiness there is in our lives."[11] Affection accompanied by an expressed concern for the affiliate's future can increase the probability of a successful interaction, the moment of truth. An organization must produce a moment in which the affiliate really thinks the organization cares and will continue to care. This affection can be demonstrated in many ways and through a variety of social media. A tweet from your organization can be warm and welcoming; Tumblr pages can be encouraging; your website can be reassuring; a comments section can be edited for positive dialogue; and a blog can be involving.

However, it takes content shaped by a common goal with guidelines and standards that promote fellowship with the community of affiliates. This does not happen by accident. Once again, we must go back to the future: training, morale, leadership, an employee sense of empowerment and purpose (all the hallmarks of a well-run organization) are critical to a savvy use of social media to achieve the goals of the organization. We must never forget that our goal is *results*. It is not unusual for Hubspot clients, for example, to have percentage gains in triple digits during their conversion from Old Media and marketing to a savvy social media approach. A Los Angeles architect, for example, saw a 600 percent increase in leads for his firm; a small resort exceeded 300 percent; and a national health services non-profit hit more than 200 percent.[12]

CONTENT POLICY—LOVE ME, LOVE MY ORGANIZATION

Content policy governs the responses of individuals to communications on behalf of the organization. Who is allowed to post on Facebook on our behalf and what are they allowed or encouraged to say with what goal in mind? What is the tone of our blogs? The YouTube postings that are linked to our Facebook page, what content do we desire? What do our tweets reflect about the organization? What are the divisions of our website and the perceptions shaped by text and graphics? How is our narrative reflected throughout all of our digital media? What type of interactive content do we want?

The preceding questions reflect the fact that effective social media must be *worked* (deliberately addressed and professionally operated) in order to attain the desired result. An organization will be unable to control every communication using social media; a case in point, is the YouTube video posted by two Domino's employees in North Carolina that we talked about before, showing the employees contaminating food during the company's launch of its improved recipes. The two were fired and charged by police, but not before the video had gone viral.[13] As the bumper sticker says, "stuff happens" (family-friendly version), but training and policy can establish preferences and limits, assisting employees in making choices. Some individuals will ignore the guidelines or, like the pair above, seek to harm to the company for a variety of reasons but, again, stuff happens and people are the primary perpetrators of *stuff*.

However, training and policy, which connect principles to practices, guide those communications toward a cumulative effect that enhances the standing of the organization with its affiliates. A simple series of questions in the following five content policy areas can assist you in building a positive and loving policy emphasis:

1. Organizational Definition
2. Target Clarity
3. Online Experience
4. Hub-and-Spoke Benefits
5. Love Handles

Let's take them one at a time.

1. ORGANIZATIONAL DEFINITION

Begin by defining your organization. Ask yourself: if my company or non-profit were a person, it would be what kind of person with what characteristics? What kind of picture, of the organization, do I want individuals in the market to carry around in their heads?

Individuals represent an organization. However, organizations tend to be thought of as "things," as entities that are often difficult to visualize. Effective content demands mental pictures, a visualization of an organization as a person with certain kinds of characteristics. Savvy social marketers describe those characteristics by creating content policies. These characteristics form the foundation of a personal relationship with the affiliates with whom individuals communicate on behalf of the organization. The organization, that has a unique and well-defined vision of itself, positions itself to create hyper-relationships with its target markets and achieve extraordinary growth.

For example, Steve Madden (the organization) mirrors the characteristics and worldview of its customers: somewhat edgy with a bit of an attitude and trendy, always trendy. As an organization, it is quintessential New York-with-an-attitude-meets-street-chic and has had category-leading growth for many years. Apparel merchandiser Steve Madden broke out of a flat-lined online presence at the start of the last decade with a website hub that mirrored its young and saucy cutting-edge women's shoes and apparel. In the non-profit sector, Liberty University (the organization) is upbeat and positive, mixing a mission that stresses Christian evangelism and a heart for God with the strengths that flow from visionary leadership and extraordinary growth to build an affiliate community that spans the globe and the kind of hyper-relationships that has made it both the largest evangelical university in the world and one of the fastest growing universities of any kind—secular or religious—in the United States. Its online social media and website hub mirror the products of this private educational and evangelical giant, offering God and growth to its affiliate community of alumni, students, family, and friends.

An organization needs to be personalized, both for its affiliates and the individual employees who communicate with the greater community. To begin the revamp of the Steve Madden website, we had to sit down with the web designers and articulate the essence of the organization as a person. Who are we? How can we, through images, text, digital application, and programming, project love and appreciation for our customers? A contact with an affiliate is, at its core, emotional, and all of the policies governing the handling of content should, at the least, result in the communication of an organization's desire for a good and positive future in all of the relationships between the individual and the organization. This trust, key to the development of a hyper-relationship, is evident in the practices of organizations that have used the digital era to enjoy explosive growth.

Consumer merchandising giant L.L. Bean, for example, has won three first-place industry awards for customer service. Its chief executive officer, Chris McCormick, attributes its digital retail domination to a culture, deliberately cultivated through training and policy, that relies on the love that "originated with L.L.'s "Golden Rule" of treating customers like human beings and our service culture has continued to evolve."[14] Above all, he explains, there is compassion for the customer, a tradition insisted upon by L.L. Bean's grandson and board chairman who said, "A lot of people have fancy things to say about customer service, but it's just a day-in, day-out, ongoing, never-ending, persevering, compassionate kind of activity."[15]

Let's call it "Affection+." It is the touchy-feely part of the equation that makes the connection, a feeling that is conveyed through the individuals working the organization's social media that he or she is doing more than a job; rather, he or she is trying to genuinely change an affiliate's life for the better through the connection of a momentary contact. Let's visit C.S. Lewis again, he was instrumental in steeling England's spine with his messages of hope, cheer, and of a God who cared during the darkest days of World War II, letting his BBC listeners know that they were intensely loved. He provided his audience, the target market segment for the British Broadcasting Corporation, with a feeling of "love [that] is not affectionate feeling, but a steady wish for the loved person's ultimate good as far as it can be obtained."[16] In the same way, the savvy social media user must train individuals to communicate on behalf of an organization to convey an underlying "steady wish for the loved" affiliate's "good as far as" the organization/affiliate relationship is relevant.

Whew! Not easy to talk about love, intimacy, and the sacrificial communication of one individual for another, striving to touch at a visceral and spiritual level, but it works.

2. TARGET CLARITY

Ask yourself, what is the target segment or segments I are trying to reach? What are the implications for strategy, especially the strategic Background and Foreground that I am constructing?

A descriptive and imaginative picture of a target is required when you are dealing with social media marketing, during which strategy sometimes drifts into ambiguity while getting lost in a torrent of conversation. The successful social marketer must accept and work with ambiguity; individual conversations often deviate from what you've deemed relevant to your creatively constructed portrait of a target. But that is as expected; conversations, especially, are one of those areas in which you learn as you go. Real life, after all, is nowhere near as clean and exact as a multiple choice exam. As the great scholar and New England Patriot football coach Bill Belichik so aptly commented, "It is what it is." And so, using the notion that all conversations serve to both formulate strategy and the source of feedback for adjustment, there must be room to research those deviations from plan and, if warranted, use that feedback in modifying your strategies. The savvy social marketer understands that, while it is critical to have a clear and creatively constructed picture of the type of person represented by the target segment, the often millions of activities engaged in each day by this class of person will contain anomalies. That's a promise, by the way, with a probability of being kept as sure as the chance that somewhere, somehow in life, you'll stub your toe. Bet on it!

Anomalies are activities (transactions, browsing, conversations, etc.) that do not "fit" the profile of the target. Keep in mind, however, that a target is a theoretical construct of the type of person that you see as a fitting customer your company, the individuals with whom a social relationship will be most productive. But prediction is not actuality and you will still be surprised at times by their choices, reactions, and interactions. It is simply in the nature of humanity to be unpredictable, to surprise with choices and astonish with behavior. The best predictor of future behavior is past behavior; hence the savvy social media user collects data on target behavior (when, how, and the amount they donate or buy, what types of offer attract the best response, etc.).

The more you systematically observe, the more effective you'll be in hitting the mark. One successful approach to bringing a target into focus is the approach known as *test-and-react*. This technique can be used to understand the reactions of targets to promotional appeals, offers, new products, and service approaches, all the *stuff* of operations and marketing in organizations. Test and react: that has been a great strength of Steve Madden and so many other successful brands. The shoe brand had the equivalent of a small cobbler shop in Long Island City, just across the East River from Manhattan, where new designs were manufactured and sent to company stores for testing and feedback from customers before being sent overseas for final production. Savvy operations and social marketing experts test products, themes, and offers on a small scale, seeking to further define and refine their understanding of the nature of the targeted segments.

Target clarity begins with a creative description of the most likely buyers, donors, and members; it proceeds from there with a process that further shifts and refines the known characteristics and behaviors of the target. Why does one communication work better than another? Which types of conversational engagement better bind target segments to the organization? Experiment, for example, with different conversation starters and adjust according to results. The most successful firms in the social era take a "dynamic approach to problem-solving" that allows them to continuously add to their understanding of the target.[17] The Hay Group described what it takes to successfully lead a digital era organization that must, to succeed, take into account the complexity of its target market segments:

> *Leadership clarifies strategies and customer needs; elicits new insights from diverse, collaborative teams; and earns the focused effort and dedication that innovation requires. . . . The Leadership creates a climate in which innovation is the expected norm, not the miraculous exception.18*

My goal is to produce leaders (yes, you, dear Reader) who will successfully navigate LAC (Life After College) due in large part to the added value of LWS (Life With Savvy_ in order to achieve the quickest and most satisfying approach to your "BHAG" (Big Hairy Audacious Goal).[19]

3. ONLINE EXPERIENCE

What are the top five online experiences that your organization desires an affiliate to have? How comfortable is the online experience for an affiliate? Does he or she feel like someone is listening? That someone cares? That the organization makes a religion out of solving problems and addressing issues, out of being welcoming and loving? These are just a few of the questions that can be asked about the social experience, which in the digital era is how perceptions of an organization and its products and services are largely formed. Through it all runs a common thread: We love you and, in fact, your needs head the list of what we want to do today. An added advantage is that love, as a strategy, tends to get talked about by affiliates; it becomes a topic of conversation both online and offline, in part, because it is so scarce in this world. Ivar Kroghrud, the chief executive officer of digital and customer consulting firm QuestBack, explains the connection:[20]

> *The age-old mantra of "actions speak louder than words" is especially true when applied to interacting with customers. It's imperative that organizations implement real-time, strategic responses to the customer insights they discover. Taking action will signal to customers that their needs are top priority. Seen through this lens, it's clear that proactive customer engagement and responsiveness will produce high-quality experiences, and delighted customers tend to share their positive experiences, both online and offline.*

Affiliates need to feel valued and loved by individual employees and processes put in place by the savvy organization. The digital world can be a hostile place where even established companies make it difficult for a customer to feel at home. Confusing websites, unresponsive customer services, and bureaucratic processes are only a few of the hurdles put in front of customers, each decreasing the chances of repeat visits, future purchases, and advocacy to friends and family. Your organization's digital world should reflect caring and individualized approaches. Save the bureaucratic and hostile for the Department of Motor Vehicles, Depraved State University administrators, airport TSA employees, and our friendly federal Transportation Security Administration (LHM—Lord Have Mercy). The feeling of being loved should be among the top social experiences for your affiliates. These experiences are summarized by the following, with each linked to a set of processes, content guidelines, and user-friendly applications:

- We care about you.
- You're the center of attention.
- We value your contributions to our organization.
- We want to make life easy.
- Your future is our concern.
- It's all about you!

4. HUB-AND-SPOKE BENEFITS

An integral part of marketing is the notion that consumers do not buy features, they buy benefits. I buy a new shop vacuum because it allows me to clean my garage more thoroughly and in less time, and *included* is a 192-piece utility set that is just plain fun, giving me at least an afternoon of involvement and play before I lose most of the pieces. I read a mystery because it provides for a personally involving and pleasurable escape from everyday life, and an excuse, in my mind at least, not to clean the kitty litter box. I watch a science fiction flick because it provides the same escape. When constructing a social media presence, we pay special attention to the benefits provided by both the *hub-and-spoke* portions of our social media presence. The benefits provided by the spoke emphasize the pleasure of immediate communication and a personal sense of accomplishment, of being able to relate to an organization, and the pleasure of experiencing that I-am-not-alone feeling that we need. The hub has all of this and more; it allows the affiliate to consummate his or her relationship with the organization, buy, donate, or join something, and assists in providing the long-term psychological and lifestyle benefits flowing from inclusion and relationship.

In the digital era, benefits are derived from the desirable attributes of (1) the products and/or services provided by an organization, (2) the experiences of communicating with the organization and other affiliates, and (3) the long-term activities engendering privileged status among affiliates in relation to the organization. The feeling of being loved and all the attending benefits is a major motivation in a socially mediated world where organizations are able to make relationship a key part of their products and services.

5. LOVE HANDLES

Love handles are not what you think (not in this context, anyway). Love handles are the content tactics that an affiliate can grasp to intensify and affirm his or her inclusion in the VRC of affiliate conversation and activity surrounding an organization. Love handles are concrete tasks and specific tools that increase loyalty to and affection for an organization. They are specific to the tool, all geared toward achieving and then intensifying hyper-relationship with affiliates. Some examples:

- Text: The use of first-person in emails from company representatives; use of the phrase "I care about you" and "You come first" in service replies; *back-at-ya* replies to praise, "and we love having you as part of our family, too!

- Graphics: No jarring colors or provocative illustrations that are not tied to relevant product and organizational characteristics; organizational policies that emphasize graphical interfaces that promote customer satisfaction rather than industry acclaim; maximum resolution for graphics based on specific social channel specifications.

- Photos: Policies that insist on professional photographs of staff and executives; deliberately posed customer service personnel with concerned expressions; customer photos, as many as possible; policies for enhancing affiliate-supplied photos; uniform product and service portrayal.

- Transaction process (Buying/donating/joining): No more than three clicks from *doing* something; reassurance built into all interactivity with the organization and its websites; feedback built into all processes, whether commercial or non-profit.

- Offers: Offers all have a call-to-action (the more specific the guidance, the more appreciative the affiliate); offers linked to activities prized by affiliates; offers substantial and rewards immediate; everyone gets something during all offers.

- Affirmation: Immediate and complimentary; thank you for emailing us; thank you for contacting us; we received your note and we just want you to know, etc.; you're special and this is how we're showing it.

THE TAKE-AWAY

Phew! We made it! It is tough talking about love, one of those touchy-feely subjects that real men and real women of the professional world avoid. Everyone *knows* that saying "I love you" should be saved for those special moments in life such as when someone allows you to take the last close parking space at the mall on the Friday after Thanksgiving or when the cable company gives you a free three-month trial for premium movie channels just when they're starting science-fiction-and-fantasy month.

It's time to open up, let the love winds blow and make love a part of our relationship with affiliates. The organization that uses social media in an affecting (loving) way is, in fact, slushy-mushy soft and, more than likely, extraordinarily effective. There are advantages to making strategies and tactics, the major influencing force in one big corporate social chick flick with tears and laughter, empathy and *love*, always love.

Savvy social media marketing relies on a process that reflects love, everywhere and all the time, from the words on a website to the quotes on Tumbler, quick comments on Twitter and Facebook, posts on YouTube, you get the picture. Love that transforms individuals on behalf of an organization is anything but haphazard. It is deliberate, the result of a myriad of policies in place while training all those in the company in relevant customer service affiliate-influencing tactics. Contrary to popular belief, love doesn't come naturally and it is not easy. Rather, it takes work, plenty of it, both in our personal and corporate lives. Social media strategies especially are highly influenced by caring/sharing approaches. The organization that takes the time to integrate this approach, train individuals within the organization, and put in place processes to monitor and improve the organization's socially mediated "love performance" will find that extraordinary growth is not only possible but probable. Love works, and as we saw earlier, even when you're a vampire. It's that simple.

(ENDNOTES)

1   http://merecslewis.blogspot.com/2010_04_01_archive.html

2   http://money.cnn.com/magazines/fortune/most-admired/2012/full_list/

3   http://www.businessdictionary.com/definition/moment-of-truth.html#ixzz1u070XOrD

4   http://www.businessweek.com/magazine/content/10_22/b4180018655478.htm

5   Top 10 Romantic Movies (pg. 2) http://www.kidzworld.com/article/5289-kidzworlds-top-ten-romantic-movies-pg-2#ixzz1wMe6Tn00

6   http://blog.hubspot.com/blog/tabid/6307/bid/31373/11-Ways-to-Give-and-Get-Customer-Love.aspx#ixzz1x3V2yN6B

7   http://www.cbsnews.com/8301-505143_162-57403971/great-customer-service-starts-with-7-letters/

8   ibid

9   http://www.skoobadesign.com/about-us

10   http://clientsfromhell.net/

11   http://csleewiswisdom.blogspot.com/2011/07/affection-is-responsible-for-nine.html

12   http://www.hubspot.com/customer-case-studies/

13   http://www.mlive.com/business/ann-arbor/index.ssf/2009/04/exdominos_employees_arrested_f.html

14   http://www.llbean.com/customerService/aboutLLBean/newsroom/stories/20100914_Chris-Shares-Secrets-Top-Notch-Customer-Service.pdf

15   ibid

16   http://merecslewis.blogspot.com/2010_04_01_archive.html

17   http://www.warc.com/LatestNews/News/EmailNews.news?ID=29809&Origin=WARCNewsEmail

18   Ibid

19   Jim Collins, Jerry I. Porras, Built to Last: Successful Habits of Visionary Companies (New York: HarperCollins, 2002) 91.

20   http://www.1to1media.com/weblog/2012/05/how_to_woo_your_customers.html?from=wk&utm_source=1to1weekly&utm_medium=H&utm_campaign=04302012

21   http://www.freshdirect.com/about/testimonial.jsp?catId=about_test

22   http://marketinghandbook.blogspot.com/2009/02/fresh-direct-makes-mouths-water.html

CHAPTER 6

TACTICS:
THINK TARGETED, DIRECTED & DELIBERATE

INTRODUCTION

I HAVE FOUND JOY IN THE MANY PLACES THAT I HAVE LIVED, PLACES THAT PEOPLE OFTEN JOKE ABOUT. TAKE NEW JERSEY. PLEASE. Okay, so not even faintly amusing. But I found joy in New Jersey, meeting some interesting people and making a difference; got religion in Pittsburgh; first guessed the weight of a hog in Iowa; and learned that there *are* polite individuals in Manhattan, although they are tourists from Georgia. What's the point? That what you find in this world depends in large part on what you are looking for. Personal or corporate, achieving your goals is, in large, part a function of defining with whom and how you're going to get there.

My advice to aspiring managers over the years has been that "you have to take people from where they are, it's that simple." Think: people are different, with differing abilities and characteristics. As the French say, "vive la différence!" Long live the difference! A bit of unexpected wisdom from a nation that Homer Simpson called "cheese-eating surrender monkeys" but, again, you have to take people from where they are. In other words, differences can be a good thing, offering shortcuts to affective strategies, understanding of the best ways to deal with groups of individuals, and more productive use of our resources. I don't go to New York and expect smiles, but delight in the occasional cheer that comes my way; I don't go to Iowa in search of bean sprouts, the hogs ate 'em long ago. And in New Jersey, I delighted in *Jerseylicious* with nary a lament about the lack of Rodeo Drive types. Want to succeed in the increasingly digital world of corporate business strategy? Don't look for Beverly Hills in New Jersey and vice versa.

Social media are used by an organization to create, shape, and manage relationships with the select individuals most critical for its success. *Select:* That's the key. In a corporate setting, social media is a tool for creating hyper-relationships that lead to more (sales, members, donations, volunteers, etc.) among a *select* group of individuals. Your organization's "basket" of social media applications and sites should, collectively, achieve the growth goals of that organization with the *selected* individuals most appropriate for its services and products. A strategic social media plan is only as good as the variety of digital tools used to identify and develop productive and profitable relationships with key customer groups, again, targets. Targets are critical market segments that allow an organization to grow and expand its product and/or service offerings in the most efficient and effective way. This chapter ties specific tools to strategy, but all in relationship to targets, which are the individuals most likely to develop into affiliates with increasingly intimate ties to an organization. Those individuals less likely are *not* targets. Obvious. Simple. Sensible. In life, simple often works.

Targeting is important. New Jersey is different from Iowa and Iowa is different from Manhattan. A Millennial (born in the mid-eighties) is different than a Baby Boomer (born in the middle of the last century) in both years and attitude. The tools that you use must fit with strategy and objectives, and all are aligned with the type of affiliate that benefits from the relationship. Using the right tool for the job seems intuitive, but it is sometimes surprising how many people forget this.

In some ways, this new digital age is similar to the Old Media age in that productive marketing requires media that fit the purposes, psychology, and lifestyle of a targeted set of individuals. The savvy social user matches the social media use profile, of the target, to the strategy and conversation. For example, if you are trying to form a community around the scrumptious candy treat, Skittles, then you do not want to advertise in, say, newspapers. This dying traditional medium has an average readership age of somewhere around a hundred years and its readers generally lack the necessary lifestyle accouterments—teeth—to enjoy the chewy candies. Instead, you use the Skittles Facebook page as your hub and use Pinterest, with its ability to graphically promote the colorful candies, to drive traffic to the hub.

In this chapter we apply the paradigm of hub-and-spoke to the selection of application and development tools. A paradigm is not just 20 cents (don't groan!) but a way of using a particular approach as a model for all other approaches. The hub-and-spoke model assumes that social media are used to drive traffic to a specific place where the activities of the organization are both apparent and made infectious, all the time producing increasingly greater engagement with specific affiliates of the organization. The loyal affiliates of an organization continually travel the hub-and-spoke, their links to the organizations strengthened through communication and transaction. Properly executed, customer relationships blossom into hyper-relationships with the organization and each other. This increases their connectivity and results in more organization-related activity and conversation over the life of the each customer.

Again, the operative word is selected. You don't marry just anyone; you marry someone whose qualities and characteristics increase the probability of a loving and intimate relationship over the years. A company will welcome all customers, but only selected customers will have the qualities that fit in such a way as to increase the chances of a long-lasting, intimate, and engaged hyper-relationship. A pool construction and supply company may sell a small pool once to an individual who decides, during a heat wave, that "it would be nice" to have somewhere to cool off after a long, hot commute. But it will have a *relationship* with a couple who love water, swimming, and water sports and want to pass this passion to their four children. The latter family will be engaged by automated online ordering of pool supplies; by rich media emails that offer water exercise accessories; blogs and apps that offer opportunity for conversation with other young families passionate about pools and water-related recreation; an interactive microsite devoted to backyard pool entertaining and lifestyles. Meanwhile, the casual-single-guy or gal will more than likely move on to other passions.

The selection and use of social media applications and technology is, perhaps, best summarized by a simple statement: "Different strokes for different folks." Enough said.

APP-Y-LICIOUS

An organization must be deliberate in constructing its network of social sites and applications, and assume that people respond differently to certain types of media. Technology should be fitted to the lifestyles and digital use profile of the targeted individual. Some individuals tweet their way from one minute to the next. Some individuals live and breathe Twitter, for example; others register, make a few half-hearted tweets, and never return. Enthusiastic Twitter users live life one tweet at a time. My Twitter feed is occasionally jammed by former students experiencing the shock of Life After College. Typical are the tweets from one young graduate beginning her career in metropolitan New York City:

- *"Still dark outside—not fair!"*
- *"Can't just go to cafeteria for breakfast—oh no"*
- *"No rest for the weary"*
- *"Nothing in fridge to eat"*
- *"Ugh, dark and cold"*

And that was just the first 20 minutes of the first day of her first job. She continued on and on: Dozens of tweets a day, documenting every part of her life for a select and patiently amused audience of friends and family. The broadcasting of as many details of her day, intimate or mundane, is its own reward for her and others like her, each the center of their universe. Conversation may be desired, but satisfaction comes with knowing that others are reading about her life, not necessarily responding. Meanwhile, others post every month or so; it depends upon interests, affinity, affiliation, and available tools. Differentiating and defining the preferred audience also points an organization toward the digital tools it needs to engage those same customers and grow their interaction with the organization. Growth in interaction takes the form of increased conversation, buying, donating, joining and other forms of engagement that ultimately lead to growth.

Groupon is a deal-of-the-day website with complimentary applications that appeal particularly to young urban professionals who are constantly on the lookout for entertainment and lifestyle bargains. The company has established itself in major urban areas as *the* place to go in search of new experiences at great prices. Its discounted gift certificates, offered on websites customized to the metropolitan areas it serves, have attracted a wide following among selected segments of the market, providing organizations that see its affiliates similarly as a way to expand their reach in the marketplace. As in most of social interactive media, its editorial content (the offers available, the types of organizations that provide the discounts, even the layout and graphics) reflects the nature of the user it is seeking. Here are some examples from Groupon's New York City website:

- Late-night hot spot serves renowned hot dogs with creative combinations of exotic & classic toppings alongside beers, shakes & cheese fries, offering $14 of food for $7
- A 60% discounted membership to an online gourmet wine club
- 40% off sophisticated Indian fast food
- Passport membership to Manhattan's trendiest fitness centers
- Up to 70% off East Beauty Spa—Relaxation is one of the elemental forces that keeps human bodies alive, alongside air, water, fire, and antifire. Ease into your element with this Groupon.

What's "antifire?" Beats me! But, then, even when I lived on the east side of Manhattan I had little interest in engaging the elemental force of relaxation, much less doing the "antifire" thing, too busy working, raising a family, and watching the Jets pursue National Football League mediocrity. On the other hand, my New York City daughter is a quintessential young urban professional and a Groupon aficionado. She lives in trendy Park Slope in Brooklyn, is an executive with a major online digital entertainment venture, and travels extensively both for business and pleasure. A surfing enthusiast, she even went to Bali on a surfing and yoga retreat. She responded to a Groupon certificate offered by a company that gives surfing lessons at a New York City beach

She is an appropriate target. Let's get into theory here (TSA—Tough Slog Alert!). Specific and defined audiences go hand-in-hand with various social media tools, becoming targets when an organization desires to use those technology and applications to promote and enhance their relationships with prospects and affiliates. An organization must choose tools that are appropriate to the characteristics of the individuals that make up its targeted segments and emphasize the benefits derived from relationship with the organization.

And now, because we live in the real world and life is not read-the-text-listen-to-me-blather-and-then-I'll-test-you, let's apply the insight:

The urban surfing school my daughter attended uses Groupon as a tool to grow its customer base by attracting the attention of young and physically adventurous New Yorkers who want to enliven their weekends with leisure 'finds' at great prices. However, because the organization's growth strategy includes increased engagement (more surfing, more lessons, buying more equipment, participating in social events such as "Faux-Hawaii Night") they need to pull in other digital applications to push these same customers into the online hub of the company: A website that allows affiliates to schedule surfing classes, buy equipment, keep in touch through social events, etc..

Principle: To build a customer base through savvy social media use, an organization must focus on the benefits of affiliation for the customer. Applied to reality: Surfing in New York City waters brings with it the benefits of attention ("Really, you're kidding! You really surf around New York?!"), an exotic test of endurance ("You must have been freezing"), and companionship (the offer was buy one lesson and get one free for a friend). Other tools that could be used as part of the surfing school's strategy to grow its customer base include Facebook (a place to talk about unique and challenging experiences), Foursquare (competitively checking in at a location and the opportunity to win prizes mayincrease the frequency of visits and keep the sponsoring organization visible to selected affiliates), and a microsite, say, devoted to winter surfing off the waters of New York City (Survivor New York: A Surf Winter on the Water). There are as many tools available as there are strategies and creative approaches. However, a *Back to the Future* approach is useful here, too.

BACK TO THE FUTURE...SIGH!!

Much of what we do, in social media, uses the tools and techniques developed in interactive direct marketing over the past few decades. Without getting too involved (this is about doing stuff, right, not just yack-yack-yacking about theories of doing something) in detail and history, the Take-Away is that we use a number of features of direct marketing to power effective social media strategies against the TERM backdrop of savvy social media use. By using digital applications in tandem with selected direct marketing concepts, we increase exponentially the power of those applications as we piece them together in a hub-and-spoke strategy framework. Those direct marketing approaches: offers, loyalty/rewards, and databases.

Classic direct and interactive marketing, even before communication went social, had grabbed an increasingly larger share of marketing communication spending because, unlike Old Media, it emphasized ROI: Return on Investment. The advantage of this over more traditional advertising and promotional communications is a refreshingly realistic worldview in which an organization asks of its efforts, "Am I getting my money's worth?" Sure, I'll pay for an expensive direct mail piece as long as it offers concrete results that justify the expense; that is, a certain percentage of those receiving the mailing will use the coupons I send out or later, go to my E-commerce site and buy using a discount code. By the end of a direct marketing campaign, I know what I'm getting for my money as customers or donors have responded directly, and I can compare results to expenditures. Results, not the squishiness of better feelings or promotion and advertising trade association awards, the staples of traditional advertising, is the refreshingly practical approach that direct marketing brought to marketing communications and has become integral to digital social media.

Results, I can't emphasize this too much. Become a person who is known for emphasizing results if you want to add value to an organization. Monetize your career, whether commercial or non-profit, by synergizing the best practices to come out of Old Media with the remarkable ability of social media to engage, grow relationships with customers, and *get results!* Stand out from the crowd, too, by not becoming so focused on technology, so in love with the next new software feature or application that you can't see the forest through the trees or the value-add through the applications (or something like that). The savvy and effective social media user understands that there is always a new application on the horizon. However, applying social technology to current prospects and customers, thereby growing the organization through a targeted and directed series of effective strategies, is what it's all about.

The savvy user views direct and interactive marketing approaches as an opportunity to inject steroids into flabby social media (Metaphor Alert!), going beyond fascination with technology and software development toys. In the real world, it is always useful to remember that somewhere, somehow, sometime, some*one* has to do something for an organization to get a return on its social marketing investment: buy, sell, donate, join, etc. So let's take a look.

OFFERS: The offer has to fit both the target and the organization. An offer is what you, the organization, promise in return for a resource exchange. In other words, you (dear donor) donate at least $100 and I (pet rescue foundation) will give you an a signed copy of the new book by our charity's founder, *Barking Up the Right Tree*; or you (dear restaurant patron) buy one entrée and get one for a friend for free on your next visit; or you (dear and beloved traveler) book your vacation now to beautiful Minot, North Dakota and receive a free dessert at the Minot Eat'n Roll Diner. You get the point, the affiliate interacts with the organization is such a way that both receive value. A relationship is monetized when an organization offers something that would be valued by an affiliate (customer, donor, member, etc.) and is calculated to prompt a transaction or exchange (donation for good feelings, membership for free month, blender for significant discount, etc.) on the part of a desired and targeted affiliate or prospect.

L.L. Bean, for example, sends an email offering 20 percent off sleeping bags and free shipping immediately before Memorial Day kicks off the summer vacation season. The Maine retail giant knows that a significant percentage of its targeted segments are outdoor enthusiasts and will respond to (1) the creative approach of the offer (a tent glowing in the dark in a field under a starry night sky as the dominant graphic in the email), and the (2) value proposition of the offer (significant savings and free postage on items that such a target) wants to buy. The merchandiser understands the nature of its customers, the characteristics that differentiate them from those who have not been selected as targets, and their desire for "getting away from it all." Consequently, they lead the target easily from the email ("all it takes is a click on this button to go directly to an awesome deal!") to the L.L. Bean website where the customer's account information is locked and loaded by technology to make the ordering process not just easier, but reassuring through customer ratings, access to customer service representatives chilling out in Maine at their warehouse, live chat, etc. Love powered by technology, benefits supported by applications, and all of it the *stuff* of engaged conversation.

FAQ 12: What social applications are best used to make offers? Wouldn't I be that much more valuable if I can walk into a company, snap off a list of the ten best applications for conveying offers, and sit back and bask in the glow of guaranteed results?!

Ahh, wish it were that easy. But life, as you'll undoubtedly find, is rarely that simple. Think you don't have the answers now? Wait until someone's pointing at you and saying, "I dunno, just figure it out!" And now you're faced with figuring it out while your husband's waiting at the restaurant, or your dog's preparing to decorate the rug because he's been locked inside your apartment all day, or your daughter has texted you that she's found a ride back from Depraved State U with a passing motorcycle gang. There are few easy answers in life, and those are generally in situations that make it so-o-o obvious that you have few options ("I'm going into labor, meet me at the hospital" or "Your mother's on the phone to talk about her colonoscopy, would you like me to tell her you're not here?"). How do you choose apps to convey offers?

The savvy answer: it depends. It begins with the target: What type of apps are they most likely to use? Age, gender, geography, and available technologies shape the answer, as does the capabilities of your organization. Do you have a crack team of graphics, marketing specialists, and room in the budget? Invest in heavy-duty email marketing software that allows attractive and intriguing email blasts. Is it just you and Buddy, the CEO's nephew? Outsource the emails and hope that Buddy knows how to read. Does the animal shelter want to offer free wellness with pet adoption? Try one of the many couponing websites. Do you have access to software developers and a cooperative head of the information technologies department? Have them develop an offer app just for your organization. Have a boss who reveals, "I only carry a cell phone for emergencies, what is this 'app' stuff anyway?" Say nothing; you'll be his or her boss within the next six months. As I said earlier, it depends.

LOYALTY/REWARDS: The notion here is that providing increasingly generous rewards and perquisites ("perks") for affiliation helps on two levels: first, the thankfulness that accrues toward an organization when it recognizes a customer who increasingly selects that organization over alternatives; and second, the innate and positive human response that seems to be a built-in when gifts and similar expressions of gratitude are offered. In the former, recognition of loyalty through frequent buyer or award processes fosters an increasingly binding relationship between an organization and its affiliates. A loyalty program tailored to the characteristics and lifestyles of targeted segments of the market binds affiliates and stimulates conversation, at the same time building for the organization a pool of high-value customers, donors, and members. In the latter, it comes down to something simple: everyone likes free stuff, from the Hollywood celebrities scarfing up 'swag' bags of free merchandise at Academy Award parties to Joe or Joan Ordinary accepting a free dessert as a "thanks for coming" to the local steakhouse.

In the digital environment, organizations large and small have the capability to create and maintain a loyalty rewards program for its best affiliates, its loyal customers. The foundation of this is a general rule of both life and business, in which a smaller pool of individuals tend to account for a larger amount of a specific activity when compared to a more general population. Studies of consumer behavior have repeatedly found that a majority of sales, for example, comes from a relatively smaller group of buyers. This is the 80/20 rule that we discussed earlier, which says that the majority of a company's sales come from a minority of its customers. The same holds true for donations and, in fact, for much of what consumer researchers would call "joining behavior." The majority of visits to a health club, for example, come from a relatively small group of members who use the facilities regularly; the majority of activities at a church involve a minority of members participating in multiple activities. The dear lady who leads the prayer breakfast on Wednesday is likely to be the Bible class instructor on Sunday and leading the fundraising for a mission on Monday. The big guy in the corner at the local pizza joint, who comes in after his weekly softball game, eats a disproportionate number of slices of pizza when compared to the total of all pizzas consumed on the nights he's there.

Loyalty, in the form of activity that helps the organization achieve its goals, should be recognized and rewarded. Loyal affiliates buy more and consume more than those who have not, or will not, reach that advanced stage of emotional involvement with a company. Encourage and engage high-frequency affiliates through attractive offers, loyalty programs, and loving—yes, that word again—conversation. This strategic approach joined with technology, produces dramatic results.

Begin with designing the loyalty and rewards strategy, including coming up with a creative approach. What are the characteristics of the target and how do we need to present our clubs and offers? Some types of activities and targets are more conducive to charitable appeals, others to edgy offers, and yet others are enlivened by humor. A climbing fitness center can reward frequent users with an Adventure Road Trip vacation, the offer communicated in bold graphics and conveying a sense of the unknown; meanwhile, a senior center can reward its loyal customers with a relaxing dinner. Different strokes for different folks, different rewards for different organizations, Daffy Duck for you, Bugs Bunny for me: It's that simple.

The same choices and opportunities are open to the savvy social media user as are offered in traditional advertising and promotions. There are reasons a particular type of person tunes you in; let the creative content of your offers and programs reflect those reasons. However, all use of technology in loyalty and reward programs, whether coupons are printed out or scanned from an app on a smartphone, communicated through a third party website or digitally pushed to a printer and delivered by snail mail, should increase rewards by frequency of activity, provide a sense of "special-ness" to repeat customers or members, and convey an underlying sense of gratitude and love with the delivery of each reward.

DATABASE: "Know your customer" in the digital age translates to "let's keep track of what affiliates are doing so we can get more of them to do more with us in the future." Again, it's that simple. No complicated theories; a database uses technology to harness selected affiliates to results.

Another given of human behavior is that the best predictor of someone doing something in the future is what he or she has done in the past. We have all seen this in our personal lives. My wife's favorite comfort food is macaroni and cheese: I can bet on it; when in need of reassurance, she turns to this cheesy delight rather than experiencing a sudden urge for bean sprouts and honey. A person who buys a black shirt is more likely to buy a black shirt in the future, while someone who walks into the workplace in a tie-die button-down with a bow-tie is more likely to wear one in the future (and be unmarried and highly likely to inhabit a single cubicle for thirty years, another issue entirely). Cheesecake in your past generally means cheesecake in your future. Anger management issues in the past; anger management issues and jail time in the future. Volunteered to ring the bell at Salvation Army in the past, and there is probably bell-ringing in your future.

The passion for collecting customers in a database began with direct marketers of the last century, who discovered that they could increase response rates by renting lists of subscribers from magazines that had audiences similar to the individuals they were targeting for their products. Their passion for results soon pushed them into compiling lists on their own of those who had responded to promotions, bought products, or expressed interest, with the resulting offers attracting increasingly more customers and achieving better results, thereby outperforming traditional advertising. As digital technologies became ubiquitous, more and more organizations established digital databases of customers and prospects, allowing more individualized and targeted marketing. At the same time, storage costs had fallen and storing information about customer characteristics (what he or she buys, age, offer responses, organizations, etc.) is among the least of an organization's expenses. It is relatively easy to collect and store information; it is difficult, however, to collect information that can make a difference in the relationship between the organization and the customer.

Savvy social media users evaluate information in light of its usefulness in enhancing relationships. Does the number of tweets correlate with buying apparel? If so, the more a person uses Twitter, the more likely he or she is to buy apparel and, besides ensuring that twitter is a spoke in a company's social network, marketers should make sure that Twitter activity is stored in the customer database for use in later promotions. Digital technologies now give us the ability to record just about all social behaviors, but not all are productive. Keep in mind that, for our purposes, an individual who participates in the major activities of the organization (buying merchandise, donating money, joining a Bible study group, taking a class) is engaging in social behavior. While commenting, conversing, and reviewing are social behaviors, so are buying, donating, and joining.

Each of these examples, involve doing something in relation to someone, and the digital online era allows us to track nearly every social behavior that has a digital recording point. In retail, for example, digital point-of-sale systems use software that transmits all purchases to a central collection point and have an almost limitless capability to link individual customers to merchandise sales. Savvy social media users keep in mind that successful use of direct and interactive marketing techniques is also critical to reaching the full potential of social media marketing. Personal details, digitally stored, form the foundation for customer service marketing, content generation, and community building; you name it, when it comes to savvy use of customer marketing. Amazon.com has led the way in making its database the core of the individual social network of each of its tens of millions of customers. Dr. Michael Mirow, consultant to some of the largest global brands in Germany, summed it up:

> In order to tap the Individualization Potential, Amazon's key value propositions to customers, besides easy-to-use browsing and searching functionalities, include reviews and content services, an online community, recommendations and personalization features. Latter comprise greeting customers by name, instant and personalized recommendations and individual notification services. Amazon provides content services,

synopses, annotations, product reviews and ratings by customers and (external) editorial staff, and interviews with authors and artists; customers can order and select gifts from the gift centre (sic) where they find recommendations from Amazon's editors. In late 1999 Amazon implemented its "Wish List" feature, allowing customers to create an online wish list of desired products that others could reference for gift-giving purposes. Since the beginning, Amazon's website included standard-setting functionalities such as customer account including multiple shipping and invoice addresses, diverse payment options, auto-login features and product preferences based on which Amazon delivers thematically matching newsletters. Users are provided shopping carts, and can merge and split orders not yet shipped; past orders are collected in an order history. Stock availability is displayed online for each product, and not yet released items can be preordered to be automatically shipped upon availability: Amazon puts "each customer at the centre (sic) of her own universe"[1]

And while this belabors the obvious, it is worth doing so: Customers are at the center of success and at the center of the social marketing and operations universe. Affiliates make the corporate world go 'round. In fact, affiliates make the non-profit world go 'round. Databases are merely collections of affiliates with attached characteristics, *data points* or variables to those who prefer jargon. Movie idea: *Social Media Transformers of the Universe*. Off to Hollywood.

CLASSES OF APPLICATIONS

So let's belabor the point. There are simply too many social media applications out there, and introduced daily, to write a definitive guide to all the weaponry in the social media armory (TMA—Tedious Metaphor Alert!). To say otherwise would be folly (unless you're the author of this book, whose readers would folly him anywhere…heh). The digital world changes so fast, with applications continually evolving that any list of tools available will be out of date by the time you read, say, the third item down on the second page of a 400 page list. Social media technology is like the weather in Texas or my oldest daughter's mind; if you don't like it, wait five minutes. It'll change. Sigh.

So how do you know what approach to take? It depends, as this suggestion by Melanie Shreffler, a business strategist who specializes in working with Millennials, points out:

> *They can use Facebook for all their social media needs, but some users are finding other networks are better for certain activities. Those other sites are becoming established niche social networks. When they want to share ideas or inspiration for a hobby, some prefer Pinterest and Twitter over Facebook to do so. When they want to update their status, Twitter beats out Facebook for nearly a quarter of Millennials. When they want to check in, for some, Foursquare is the best way to do so. Perhaps most interesting are the actions for which Millennials say all social sites are about the same; those represent opportunities for the next niche site. This chart shows the preferred networks for a variety of actions for which Millennials use social media . . . Marketers who want to use the*

social sphere to connect with Millennials should consider how teens and 20-somethings use each site and tap into those natural behaviors. Planning a campaign on Pinterest should reflect users' creativity and desire for inspiration. A campaign on Tumblr should be informal and give users some fun eye candy tied to a theme.[2]

The bottom line is this: You want something to happen—someone to buy, sell, donate, or join—then begin with the differentiating characteristics of the target and work back from there. Understand the content of the conversations they want to share and their receptivity to specific creative approaches and activities. Once you've done this you may now match the psychological and behavioral functions of the target with appropriate applications. Nothing replaces an intimate knowledge of an organization's targeted segment(s), knowing who they are and how they communicate:

- Are you a business-to-business vendor who wants to engage a member of Congress? Twitter would be an effective spoke. How do I know? Because 82 percent of members of the U.S. Congress have Twitter accounts[3]

- You run a Boy Scout Troop. A combination of graphics and text on Tumblr would offer a specialized blog that would combine text and graphics. Your scouts could post merit badge photos, community project graphics, etc.

Use the apps that are productive for the type of organization you are. Construct a market basket of tool and techniques, apps and software bundles using a hub-and-spoke structure, always careful to deliberately choose applications and development tools that "fit" your best customers.

Keep it targeted. Everyone may use social media, but specific individuals have a social media use profile that fits your organization. You construct a social media network for your organization for a *specific* target segment (or segments) and an understanding of the operational opportunities and constraints of your organization. The popular DailyCandy website spends millions on email software maintenance and development, but their primary means of communication and networking with their customers *is* email. Makes sense! But you are an independent tavern operator in White Sulphur Springs, West Virginia, where a relatively inexpensive add-on to your standard personal email application would allow you to reach your engaged segment of a total market of less than a thousand residents. Again, a savvy and definitive approach to reaching your target can best be described in this way, it depends!

Here are some broad guidelines:

WEBSITES: Websites are the central place where the activities are performed. Sometimes there will be a separate corporate site aimed at investors, but generally savvy companies have a main online place that offers investors a flavor of the company while providing a digital hub for the core activities of the organization. The site should be created with the customer or affiliate in mind; that is, easy to navigate, reassuring, displaying the same qualities that are unique to the organization and differentiates it from competitors.

MICROSITES: These are websites often used for a one-time campaign, issue, or product. Wikipedia describes microsites this way:

A microsite is an Internet web design term referring to an individual web page or a small cluster (around 1 to 7) of pages which are meant to function as a discreet entity within an existing website or to complement an offline activity. The microsite's main landing page most likely has its own domain name or subdomain.[4]

Coca-Cola has a long-running and strategically central microsite in its My Coke Rewards website. This microsite has awarded millions of prizes to Coke customers since its beginning in 2006 and has become a must-stop site for Coke customers. In addition, global brand has more than a half dozen other microsites, each catering to a more specialized Coke affiliate such as investors, bottlers, and other loyalty programs, and collectively engaging Coke affiliates.

NEIGHBORHOOD APPS: Mr. Rogers knew what he was doing when he sang "It's a beautiful day in this neighborhood." Neighborhood Apps are geography-dependent applications that rely on the experience of customers in specific geographic locations. Foursquare, for example, combines rewards with check-ins with a mobile application at specific locations, while it also features a variety of software features that allow users to customize and add value to their experience at specific places.

Other Neighborhood Apps include entertainment and travel review applications, which provide information to users who may want to dine at a particular restaurant, stay at a hotel in a specific area, or participate in a variety of events and attractions. Why include these in the Neighborhood Apps category? Because ever-changing applications such as Yelp, Hotels.com, Urbanspoon, TripAdvisor are geography-based, offering utility to users when they apply the technology to shaping their visits to specific geographical locations. They use social technology to make dinner reservations, check traffic, schedule flights, determine the value of special events, etc. At any given point there are dozens of feature-filled applications in the Geographic Apps space, all offering a differential advantage for savvy social media users. And, of course, if your organization has the resources, you can have customized Neighborhood software developed for your website.

CORE APPS: Core apps are one-of-a-kind applications that can be used in a variety of ways in the hub-and-spoke process but are committed to providing robust growth for organizations seeking to use them for business. Core Apps typically have a central website and complimentary applications that, in a sense, collect customers while providing a multitude of ways for businesses to connect with those customers. Core Apps also make available a variety of applications that, on the one hand, increases the enjoyment and pull of the site for users while, on the other, offering technology that provides business users with a multitude of ways to leverage the power of the individuals the site has collected. Core Apps provide *hub* features that bind the social network of the company to affiliates. Core Apps may include, for example, concierge software in the hospitality industry that allows hotels to download dining and activity suggestions to guests upon check-in at the front desk.

Facebook may be used as either a part of the hub of an organization's social network or as the hub itself— that's how powerful it is as a Core App, with more than 30 percent of all global Internet traffic (and growing)— and/or as a powerful spoke driving traffic to the hub website of a company. In addition, it has a robust and growing collection of development tools for organizations to use in building individual social networks. Useful Core Apps are usually designed with ease of use in mind for ordinary consumers and facilitating engagement. An example of this is the Coke Facebook page, which was started by two affiliates who simply wanted everyone to know about their love of Coke. The international brand has "long attempted to foster meaningful emotional bonds with its customers, now living in over 200 countries worldwide."[5] Its chief executive officer paints a picture of a company that has deliberately and strategically tied itself to its affiliates, shaping itself as "much more than just the product. It's about universal refreshment, about moments of happiness. It's not that Coca-Cola represents the American flag. It's a unique representation of optimism."[6] Coca Cola has more than 33 million fans of its Facebook page.

In the political world, for example, Sarah Palin has used her Facebook pages to shortcut the mainstream media, which has been vehemently opposed to her outspoken, independent brand of conservatism. She has gotten the word out about issues and candidates important to her, featuring promotional highlights and position papers through an interface that reflects her no-nonsense and no-frills approach to politics. Her Facebook pages also contain a link to her SarahPAC website, where Palin fans can join and donate to her political action committee.

On the *spoke* side, Twitter has proven especially useful in driving traffic for a variety of commercial, non-profit, and political organizations. As a *spoke* Core App, it allows affiliates to express themselves creatively, all the while, providing organizations with a place to draw attention to their various activities. Organizations use it for the same reason Willie Sutton gave for robbing banks, "Because that's where the money is." Organizations use Twitter because, as Bloomberg BusinessWeek put it: "Companies are joining Twitter for the same reason politicians attend the funerals of famous people: It's where they can find their constituents and hold close, informal conversations with them."[7]

But regardless of the application, savvy corporate users are careful to take a conversational tone that builds their brand and sets a perceptual tone for the organization. Starbucks uses its *hub-and-spoke* social network as, with its Twitter presence, a natural extension of its chief executive officer's public statement about its corporate soul "Starbucks is more about people than coffee."[8]

The social media lineup changes every day. Pinterest, for example, has been gaining speed. Pinterest now beats Twitter in referral traffic.[9] But it is sending a different type of customer, more likely to be rural, to have children, be female, and come from a household with more than $100,000 in income.[10] Think target, think specific individuals, and matching the users of any individual social medium with the existing and potential affiliates of your organization.

The best approach to choosing tactical tools and applications is to see what fits. An app, like a company for its customers, has a set of features and benefits for an organization as it constructs a strategy to achieve its growth goals. The savvy organization constructs a social network with its corporate objectives and preferred target individuals in mind. Once that is set, it then looks at its resources and, accordingly, selects social media to form the foundations of its programs. As complex as an organization is, as many moving parts that it has to fit together to create and maintain a credible approach to corporate growth through socially mediated affiliates, one piece of advice still applies: Keep it simple. Social media selection is best approached with a heaping helping of common sense. The Los Angeles Times has surveyed the popular social media and came up with a simple way of looking at them, "Social Media Explained With Donuts."[11] The following chart is from the article:

SOCIAL MEDIA EXPLAINED*

Twitter	I'm eating a #donut
Facebook	I like donuts
Foursquare	This is where I eat donuts
Instagram	Here's a vintage photo of my donut
YouTube	Here I am eating a donut
Linked In	My skills include donut eating
Pinterest	Here's a donut recipe
Last FM	Now listening to "Donuts"
Google +	I'm a Google employee who eats donuts
Tumblr	This is a really beautiful photo of a high-end donut
Reddit	There is a conspiracy in the donut industry
4Chan	Donut. Hole. LOL.

*Adapted from Andrew Simmons, "Social Media Explained with Donuts," LAWeekly, Feb. 7, 2012

Social media applications are complex but simple. There are so many types of technology, so many different approaches and features that an organization risks getting in over its head unless it keeps in mind a single goal: Grow a relationship with a specific type of affiliate whose engagement with the company will lead to corporate growth.

THE TAKE-AWAY

It depends. Even with the selection of technologies on which you can base your social media strategies, the target is everything. It all starts with the individuals who are your most engaged affiliates. When you prospect, you're looking for more people like your most loyal and engaged affiliates and deliberately shape your digital network through content, application and conversation to attract those types. When you offer specials and deals, you're thinking about what it takes to get targeted prospects and existing affiliates to buy, join, or donate. When you select social media, you keep in mind the types of applications that they find natural to use for self-expression and making sense of the world around them. And when you build a website, you acquire the tools and software with features that fit with your best customers.

Applications and other technology should be chosen with selected individuals and the resources of the organization in mind. Even when a technology is appropriate for a targeted group of affiliates, an organization may not have the skills and internal strengths to take advantage of it. So a savvy social user is also a deliberate operator, taking care to assess the strengths and weaknesses of his or her organization. In this case, a less feature-rich technology may be preferable until the organization acquires the skills needed to fold into the operations of the organization

Start small. Work smart. Let your target drive the technology. And eat your vegetables. Simple advice for the savvy social operator.

(ENDNOTES)

1   Michael Mirow, "Seminar Paper on Strategies to Achieve Market Leadership: The Example of Amazon," Summer 2005, Berlin May 17, 2005 Presented at Technische Universitat Berlin

2   http://www.mediapost.com/publications/article/172404/millennials-are-leading-the-social-media-explosion.html#ixzz1us0I9bz2

3   http://www.politico.com/news/stories/0911/64689.html

4   http://www.postclickmarketing.com/2011/03/30/3-factors-in-using-a-microsite-vs-landing-page/

5   http://www.warc.com/News/PrintNewsItem.aspx?ID=28854

6   http://www.warc.com/LatestNews/News/EmailNews.news?ID=28854&Origin=WARCNewsEmail

7   http://www.businessweek.com/managing/content/oct2009/ca2009106_370257.htm

8   http://images.businessweek.com/ss/09/10/1006_twitterville/9.htm

9   http://techcrunch.com/2012/03/08/pinterest-now-generates-more-referral-traffic-than-twitter-study/

10 http://www.modea.com/images/modea/site/uploads/logos/pinterest_infographic.jpg

11 http://blogs.laweekly.com/squidink/2012/02/social_media_explained_with_do.php

CHAPTER 7

BUILDING GROUPS: IN COMMUNITY WE TRUST

INTRODUCTION

LOYALTY, IT IS A TREASURED AND RARE CHARACTERISTIC THESE DAYS, finding individuals who

are faithful to their principles and commitments. Corporate growth, however, relies upon customer loyalty to drive growth, the result of a process that has brought the organization and its affiliates together. However, a savvy social organization naturally builds loyalty and enjoys the many rewards that go with it; increased affiliation, of course, followed by active advocacy on the part of customers and an outsized return on the social networking investment. Loyalty is as much a part of technology-driven marketing strategy as jargon and timelines.

An organization builds community, which in turn engenders loyalty. We group individuals together into a network of individuals with shared characteristics and behaviors that exert a powerful pull on individuals. For organizations that build a coherent and positive community around its services, products, and brand, that power drives growth along the paths blazed by the totality of the social network shaped by the company. As an organization establishes itself as an emotional touchpoint in the lives of affiliates, the loyalty of the latter grows and pulls others into the community of increased purchases and activity. A growing community of brand loyalists pulls less attached affiliates toward more attachment and, consequently, increased loyalty. An increased base of loyalists sets the stage for widespread advocacy of a company and its products in a community. This is desirable, but much depends upon the nature of the community that's been created. A community, by its very nature, pulls individuals within it to or pushes them from certain characteristics and values. In other words, individuals tend to conform to the groups around them, acquiring the characteristics of the community.

A carefully constructed and positively nourished corporate community, which links individuals with others along a common and characteristic set of activities and worldview, uplifts individuals toward hyper-relationship with the company. The savvy social organization strategically builds a community defined by its most loyal affiliates (customers, donors, or members) knowing that individuals entering the community are pulled toward the norm. The savvy social company uses the ubiquity of social media to first sweep the skies for individuals to enfold in community with the organization and its affiliates and then once there, holds them tight, binding them in a Virtual Relationship Cloud (remember VRC?). It encourages affiliation that, with generous scoops of love and emphasis on benefit-driven communications and marketing, turns into extraordinary satisfaction, identification, and loyalty. And loyalty in the social media age tends to lock the individual into a place in community, which then supports increased intensity as displayed in advocacy and activity (buying, donating, etc.).

This chapter links social media to loyalty marketing and the building of community. The savvy organization takes care to select specific types of affiliates (this individual will donate more, this shopper will buy more, this individual will enroll in our classes) and then uses the glue of social media to bind them together in conversation. That conversation is both an outcome and display of community, in which two or more individuals bond over the vast and strategically directed possibilities for conversation around an organization.

The savvy company builds a bond of trust between the affiliate group on the one hand, and the organization on the other. Social media are used to both create and manage groups of high-frequency and high-value customers, both in commercial and non-profit organizations. In this chapter, we provide you with an understanding of the power of group creation linked to continuing social media conversations and content. This chapter will outline a variety of ways you can encourage interaction, focus on the proper transactions, and have appropriate creative content, as evidenced both in social media conversations, promotions, and advertising. We also note that social media uses the same approach used by Old Media advertising in that creative content attuned to the lifestyle and characteristics of the community is a key to building loyalty. The savvy social organization needs *stuff* (there we go again, just LOVE that word!) to talk about, and that stuff, all related to the organization and its community, should be involving for the type of individual bound to its community.

We are social creatures and the age of social media has intensified our social-ness (NWA—New Word Alert!), allowing organizations to get in on the intensity that has largely been available only to individuals.

American Dairy Association: "Behold the power of cheese!"

Savvy social media user: "Behold the power of social media!"

COMMUNITY IS AS COMMUNITY IS BUILT

Savvy use of social media produces groups of involved affiliates in community, their conversations critical to the continued role of individuals within the community. Those within the community are extraordinarily connected, both with the organization and other individuals within that community. The community constructed by Coca-Cola, one of the world's premiere brands, is a case in point.

My wife and I met in Atlanta and periodically visit, the last time taking in its world-class Georgia Aquarium. I mentioned to a graduate student that I had visited the area, but did not get a chance to visit the adjacent World of Coca-Cola, her eyes widened.

"I love Coke!" she gushed, "I don't understand how you can be right outside and not visit them." She then described driving to Florida with friends during a semester break and insisting they time the drive through Atlanta to coincide with the hours when the exhibit was open. She loves Coke and all it stands for. She is a member of their site devoted to providing prizes to the Coke community, My Coke Rewards. This is where the individuals in the community can "reward yourself or others" and "get a little something to brighten your day," "win big," "try your luck," or, for the more social minded, donate to a school and "help students become active social learners."[1] She frequently visits the Coca-Cola Facebook pages, along with 33 million other fans of the world's largest brand. The Coke organization is expert at building hyper-relationship. Commpro.biz gives us the why and how of the effectiveness of the global soft drink brand:

> Creating a strong social media presence takes a great deal of work, intentionality and strategic focus. Coca-Cola has always been a social brand—one that is focused on their customers and building a sense of loyalty and community. Facebook is but an extension for this brand that already understands the importance of engagement with their customers. They are global in their approach and fully integrated. Even their Twitter feed and YouTube channel highlight their commitment as a global brand....Social is not

something new for this brand. If the belief is there and the values are there, then the team of individuals who are responsible for implementing the social interaction opportunities across all properties (Facebook, YouTube, Flickr, Twitter) will be able to fully capitalize on the opportunities that communication technologies provide. It's not rocket science for Coca-Cola, and it's not work—it just comes naturally.

Done right, building a community should come "naturally." It is what we, as social humans, do, joining ourselves with others from fans of the Jets of the National Football League to Harvard University alumni. We like to identify, prefer to identify and, once again, savvy social marketers take advantage of our tendency to bind with others. This student is part of a community that, in part, derives its identity from Coca-Cola. While Coca-Cola is not her life (she has one and a very active one at that) it is a significant part of her identity. She converses about Coke, has Coke memorabilia, and is disappointed when she happens to land at a Pepsi-products-only restaurant. At the same time—and even more importantly for Coke—she has become a Coke evangelist, one of the people who derive such pleasure from an organization and its products that she urges others to join with her, not because she wants to sell them something or spend their money (which will happen if they choose to identify with her community), but because she wants them to benefit from the relationship. It is a deep and satisfying relationship, and she wants others join her in deep satisfaction.

She and others like her are part of the reason why visitors to the Coke pages continue to grow, producing "strong fan engagement, typically running seven posts a month, each one of which garners more than 235 comments and nearly 1,750 'likes.'"[2] The Atlanta company has deliberately engaged its fans through its social marketing strategy, working with them and including them in building, for one, Facebook fan pages and putting in place the *spoke* media that drive customers to its hub websites.

The strategy is deliberate: work with and reward affiliates. Talk to them. Build a community by engaging the individuals, the affiliates who identify or, based on their characteristics, are likely to identify with the emotions stirred by the brand. And, once included, keep them included by stimulating conversation based on interaction. How? Keep offering rewards, reward them for purchases, assist them in thinking about the company and its products with games and contests, and offer the opportunity for them to surround themselves with artifacts (Coke serving trays, dolls, vending machine replicas, vintage ads) that are emblematic of the community and its attachments.

Relationships need *stuff*, something to talk about, rewards for loyalty, expressions of love and relationship and affinity. You get the picture. A savvy social media user builds affiliate relationships that are similar to a good marriage. A good marriage relationship is not just talking about feelings; rather, it is talking about the causes of feelings, of activities and preferences, of what you've done separately and together, what you want where you're going, and what you've wanted where you've been. It is talking about doing and the effects of doing, both experiential and psychological. In our personal life, we don't just have a relationship; we have *stuff* that is evidence of our relationship. A company/affiliate relationship in a savvy social network is a good, perhaps great marriage.

A relationship doesn't just happen without activity, things to do, conversation, and transactions. I love my wife, yes, but what does that mean? It means that I may hold her hand, whisper sweet nothings in her ear ("How about getting me an iced tea while you're up?"), and stick with her through thick and thin ("I love you no matter what you look like"—not an inspiring comment, but…well, I tried). The same process applies to loyalty to an object, such as a Honda Accord. You don't just drive it; you talk about its styling, its durability, tell an anecdote about how comfortable it is, and reliable in traffic and you intend to buy another when the time comes. Loyalty: You're now part of the Honda owner community.

Community is the basis of conversation. And conversation about community relies both upon the benefits to the individuals and the features that fit those benefits. The student community is not just a group of individuals in one place taking courses. The community of students is those who share the experience of taking courses and discussing them, passing judgment on the instructors, and going to football games, concerts, perhaps sharing three hours of conversations at Starbucks. That's community.

Organizations must keep in mind that conversations are rooted in *stuff*, things you do or about which you think or feel. You experience and/or acquire the stuff of life and then talk about it. To stimulate community, there must be stuff—experiences, products, activities—and shared commonalities. Conversation in cyberspace, one person speaking to another and then speaking to another, create bonds and thereby, community.

Critical elements in the building of community for an organization through conversation include: providing affiliates with **interaction**; appropriate **creative** content in offers and blogs customized to the characteristics of the target segment(s); and **transactions**, the capability to buy, sell, join, and donate. Conversations revolving around the central functions on which your company operation is organized. Let's take a look:

1. INTERACTION.

You need to provide *stuff* for affiliates to do. Coca-Cola not only offers products, but it makes a game with rewards out of showing how many products you have consumed. Games, contests, prizes, puzzles, and downloads, the online world is filled with ways to involve individuals with an organization by *doing* something. "Check here," "click this," "'like' that," all have the advantage of providing stuff to do. Once you do it, you talk about it, becoming even more invested and, in short order, slide into the more intense and pleasurable hyper-activity in relation to the organization that presages loyalty. When you reach the loyalty stage, the experiences are so pleasant that they become remarkable, something worth remarking upon, defined as something worthy of comment and a major driver of conversation:

- You wouldn't believe the Coke Rewards I got, a three-day stay in Myrtle Beach!
- They're giving a free annotated Bible for everyone who goes to Pastor McSnooze's Bible study, how about coming with me next week?
- Hey, if I apply to Depraved State University for early acceptance, they'll give me my own DSU Slacker Kit!
- Goodwill is having this awesome promotion; for every item you donate, you're allowed to browse the store and get a comparable item in the same category.
- It is so easy to donate to our South American missions group; you can sign up for your own page, set goals, and see where your money is going. They even have an animated mariachi band that plays every time you make a contribution.

Providing online activities worth commenting upon is a required ingredient for successful marketing in a social and digital age. One of the savviest social media users of all time is Spotify. Spotify is a music streaming site that is on its way to evolving from a significant presence in the music world to a dominant global entertainment presence. One of the more clever strategies it has is allowing the open development of applications for its site, providing users with new things to do with the site, and a broad range of functions that assist in intensifying the music listening experience.

The Spotify website is a hub, allowing Facebook and Twitter and a variety of smartphone applications to feed into the site and vice versa. Spotify developers have achieved seamless integration with a growing host of other applications that allow users a variety of options in letting each other know, in their particular music community within the overall Spotify community, what they're listening to and the new songs they've discovered. As Spotify software design executive Sten Garmark noted, Spotify could become the "OS (operating system) of music" and, perhaps, a fully functional social platform of its own.[3] The advantage of this for community is an endless series of all-things-music to interact with, as both music labels and the industries that have sprung up around them provide Spotify affiliates with material for listening and affiliate play. This is the result of savvy social operations: social community, social play, and growing enthusiasm, affection, and loyalty. As XOM Reviews, a trade website, devoted to music characterized it: "Because music is social, Spotify allows you to share songs and playlists with friends, and even work together on collaborative playlists, Friday afternoon in the office might never be the same again!"[4]

"The Future's So Bright, I Gotta Wear Shades" is not just a song on Spotify; it is an apt description of the ability of the Spotify management team to monetize the emerging trends in social media. Social marketing is the foundation of Spotify technology and operations.

2. CREATIVE

Shhh. Don't tell anyone, but writing counts, as does graphics and images and all of the parts that comprise content. The individuals in a community are, by definition, related to each other in some way, shape, or form, and in many areas see the world in similar ways. They can, therefore, be counted on to react in similar fashion to online communications. This is why it is important to have content that is creative in keeping with the nature of the community.

The nature of the community demands content that shares creative and unique characteristics. The stronger the communities, the more telling are the characteristics that set them apart from others not in the community. The community surrounding the San Diego Comic-Con International is known for the colorful nature of the sell-out crowds it attracts every year. Comic-Con is known for its unique characteristics. What once was a small comic book convention has evolved into a pop-culture expo that ranks as one of the largest conventions in America.

This community around this convention is marked by a fervent, at times fevered social network that includes blogs featuring action figures and media projects galore, as well as a film that "explores this amazing cultural phenomenon by following the lives of five attendees as they descend upon the ultimate geek mecca."[5] It is not just any creative content that captures the attention of the members of this community, the content must appeal to the kind of adult who thinks nothing of dressing up in a She-Ra, Princess of Power costume and walking the streets of a major American city. Obviously, this is not your everyday garden-variety individual, as shown by the following description by a reporter for the Orange County Register attending one of the conventions:

Back in the halls outside the convention floor, costumed fans pose for photos, the more elaborate or unique the costume the more requests one gets. Karl Zingheim of San Diego came as a centaur, with a six-foot-long horse's, um, tail quarters attached to him with an elaborate piece of homemade machinery "The whole point of cosplay (costume play) is seeing and being seen," he says. "So if you've got an over-the-top costume, people are going to notice you." Meanwhile, Jennifer Peurifoy, a 31-year-old mother of two from San Diego, arrived in pink from head to toe as Jem from the 1980s animated TV series "Jem and the Holograms." She loves Comic-Con, her husband Chris? Not so much. "He does not like Comic-Con," Peurifoy says, laughing. "He thinks I'm a dork, but he loves me anyway."6

This is community at its best (or worst, depending upon your point of view). Dozens, perhaps more than a hundred blogs and social media sites have sprung up around this popular culture phenomenon. The social network has a unique flavor and contributes to the identification of individuals who are part of it with various popular culture characters from movies, comics, television, and graphic books. Their connection is intense and reflected in the creative; the manner in which content is implemented through graphics, audio, text, and video techniques.

An organization selects an appropriate target; that target, in turn, shapes the nature of the messages used to involve and bind individuals to the community. The creative content then shapes both the relationship and the perceptions of the community, both from the inside and outside; and creative then becomes an integral part of the nature of affiliation with and attraction of others to that community. The creative produced by the organization and that produced in conversation (BTW, graphics are an integral part of the conversation) by affiliates are all of one piece. For example, a target of young women with an attitude is integral to the branding of apparel merchandiser Steve Madden, which demands irreverent creative that plays on the self-perceived "cool" of the community. Consider this creative—the promotional wording—on an official Steve Madden webzine:

Nailing Hollywood's celebrity nail artist **Stephanie Stone** gets inspired by our STUDLY for this awesome mismatched leopard (sic) and gold mani-pedi design. Also, her Tumblr is like RAWRR. "The moment I saw this shoe, I knew I wanted to create a nail look inspired by it! I wanted to design a nail that mirrored the shoe's bold style. The black accent nails compliment the animal print and ensure that the overall look isn't too match-y. I finished the nails with an elegant gold tip as an ode to the shoe's gold studs." - *Stephanie Stone*[7]

And then affiliates respond in creative kind, typified by these posts to the Madden Facebook page:

So I buy so many steve madden shoes but there are a lot(sic) more I want and you guys don't make them in 11 and if u do u don't make a lot(sic) why is that? And will u guys ever change that. Because I bought a few wedges last year because u had some 11 but this time sound I only c 2 typed of wedges in that size

If any of you are thinking about getting these,don't hesitate! Their (sic) comfy,& the leather is a really soft,good quality leather. My boyfriend ⌧'s them! Plus,Steve Madden always has ur size no matter what,& always has coupons & sales!

New England Patriots coach Bill Belichik is known both for his savvy approach to football and a philosophical bent. Belichik, his words dripping with discernment, summarized the essence of this conversational combination of affiliate and organization when he uttered, in response to a question about what another team was doing to prepare for the Patriots, "I don't really care about that. I'm concerned about what our team does. We all have a job to do here."

"We all have a job to do here." In the age of social media, a company is defined 24/7 by both the company and its customers through socially networked conversation about the organization and its products. Both customer and company "have a job to do here." Old Media, Old Marketing: Our company has a job to do here. Social Media, Conversational Marketing: Both company and customers have jobs to do here.

The nature of the job is defined by the characteristics of the brand. Brooks Brothers customers aren't Vogue-flashy, they're crisp and professional and convey this image in digital conversations. No odes to, say, a "Steve Madden 4 the Cool People" wedge shoe or paroxysms of gush; that's not the kind of affiliate community they are, and the company they are in hyper-relationship with matches their buttoned-down demeanor. Its Facebook page asks, "Alright gentlemen (and we expect honesty), can you pull off a bow tie or "knot?" What's the secret (other than it being from BB of course)?" And a woman, young, professional, and put together comments, "I love that tunic top!" The savvy social organization and its affiliates are creatively synchronized, and at the same time distinctive from others not in community.

Creative reflects the commonalities between community and affiliate. The agency that produced a promotional campaign for a recent Mazda automobile, targeted younger and edgier males. Because it was focused on initiating conversation, the company deliberately fit the lifestyles and psychology of its target market segment to the social creative of its campaign. In its creative content, it combined the use of a professional surfer and a contest asking viewers to submit for personal stories of achievement. The creative gave the target segment (the community) guidelines and assistance while asking them to tell their own stories and have a chance to win prizes. The social media campaign worked, and the lesson for the agency was this:

> *A social campaign's shelf-life is largely dictated by its content. While most businesses today have a social media page with some content thrown in, campaigns sometimes fall short of expectations due to lack of foundation in the form of goals. Creating promotional content isn't simple; yet, it's not rocket-science either. Less sales talk, more storytelling, engagement and listening with a dash of contests thrown in is the perfect recipe for dishing out compelling content.8*

Compelling content: That's a requirement for success in building the trust of a community. Don't waste the time of your affiliates, who generally know when an organization is just phoning it in. Savvy creative is propelled by a synthesizing and involving idea pulled from the nature of the target. In advertising, this is known as the "big idea." This is the creative idea behind the social media content, behind the interaction that makes the interaction memorable and gets the attention of affiliates. The more in tune with the affiliates the idea is, the more compelling it is and more people it will engage. Advertising and graphics blogger David Engel says of well-targeted creative content "Big ideas are fresh and provoking ideas that hold a viewer's attention. They stimulate the mind and–many times–stir the emotions. Big ideas are simple and easy to understand. They are not lists of benefits."[9] So the big idea both propels and compels; the former pushes forward the narrative while the latter compels affiliates to engage the content. It is a creative thing.   Here is another example from one international agency in the advertising arena:

> *The creative team, led by Mark Figliulo, CCO (Chief Creative Officer) of TBWA/Chiat/Day (the United States division of TBWA Worldwide marketing agency) in New York, devised a campaign that positioned the brand as an invaluable tool to minimize cravings and withdrawal symptoms. The work used direct language and a sense of humor to show how Nicorette can help. TBWA introduced the "Suckometer," a fictional device that measures the "suck level" of a smoker's craving. In one commercial, the device, placed in the passenger seat next to a driver, beeps with a red-light warning when the motorist catches sight of someone smoking. In another, for mini lozenges, a man can only focus on his cravings even though there is a shark attacking his arm.10*

The big idea gets attention, engages the affiliate, and provokes more conversation with attending interaction. It is a single animating idea, expressed in a single statement used to link and drive the narrative of various tactics, portions, whether through Facebook, Twitter, Tumblr, Pinterest, or a host of other applications in existence that comprise a social network, or the multitude of digital applications that will go online in the weeks, months, and years to come The big ideas has an element of warmth, of something personal that gives the affiliate a reason to engage.

3. TRANSACTION

A community revolves around transactions. Transactions stimulate conversation, and conversation invites community. Every transaction either achieves something or it does not; the blender I just purchased at Bed, Bath, & Beyond can do everything but drive my car, and I also received 20 percent off with a coupon that had earlier been mailed to my home. Not only that, an associate in the store gave me some tips on using it to make soups, directed me to the company's website for related equipment, and then guided me to a link to a website with soup recipes. All in all, it was a morning of love shown by the associate who cared enough to assist a customer, by an individual working for the retailer who put me in touch with like-minded individuals, and now having the means—a Ninja Blender—to become a fully-functioning member of the wider gazpacho and hearty soup community, by the blender merchandiser who knew exactly what to design for someone like me. As I said, love. In response to all that love, I told my family and friends about my morning; I mentioned it in an email to a former colleague when setting up a lunch in Manhattan; I used it as an example in this book. My transaction provoked a wider set of discussion and additional transactions, of added love to the virtual cloud surrounding Bed, Bath and Beyond, of a greater intensity of friendship as that former colleague bought a Ninja Blender ("Awesome, it chops up everything, including my cat when I get too annoyed…"). Transaction to conversation and back: Social Media, the Final Frontier!

····· **FAQ 13:** How do I use this insight into creative, targeted content in my career?

Great question, the answer: Always ask yourself, "Am I getting stuff done? Does what I'm doing result in activity relevant to the growth of my company?" Your value for a company directly correlates with your ability to deliver results for your organization. And results are desirable transactions that drive toward the company's growth goals. Let's say you are hired as a marketing manager for an industry trade group, as one of my students had been, with responsibility for creating a social media capability for the organization. First question to ask: What do favorable results look like? For this organization, more people attending conferences, more companies moving up to premiere membership, more sales of industry publications, etc. In two words, it's "more transactions." My former student's new employers were quick to recognize and appreciate his linking of online conversation and activity to concrete results for the organization, as expressed in transactions. Their appreciation, however, was expressed in expanded responsibility and increased compensation. Now, let's bring it back to you: Clear the fuzziness of higher education from your brain and stride boldly into the marketplace, insisting as much as possible on measurable results for your employers. You'll be appreciated and, ultimately, rewarded (if not by a present employer, then by the next. Loyalty is as loyalty does, you know).

Savvy social media use, at some point, involves transactions, satisfaction, and benefits. Let's go back to the graduate student who is a fan of Coke and a frequent visitor to the My Coke Rewards site. Part of her discussion of the site (the stuff that builds community) is the transactions that result.

- "They are re-introducing Cherry Coke in some markets and gave me a coupon for free drinks"
- "I saved so many caps that I got a free trip to Disneyworld!"
- "You wouldn't believe this waitress, she's a Coke fan too and didn't charge a fellow Coke-head!"

Consumerism (the buying of consumer goods and services) is as natural to life as breathing. Consumption is what we do, and consumption is what you find as you work through a chain of conversation. Start a blog about love? Fine, but love is far from an abstraction; rather, it is specific types of conversation held while consuming products: gentle conversation, for example, in an Italian restaurant, pasta and sinful tiramisu.

Transactions yield concrete results, and concrete results are the stuff of conversation and growing affinity. A key part of forming a social community is discussing results: who gets the most mileage from his or her hybrid Chevy Volt before the battery bursts into flames; passing a licensing exam for beauticians; , this week's expected Jets loss (sadly, more often that I'd like as I'm part of the Jets fan community), etc. Transactions are inseparable from results. When there is a favorable result for both parties, it is generally referred to as a win/win (yes, trite but true); I give you a $10 donation for your mission trip, you give me a chocolate-chip cookie and the affirmation that I'm doing right by my faith community. When it only produces favorable results for one party to the transaction, that's a win/lose and, as another great philosopher, Bob Dylan, suggested, "Pick up your money and pack up your tent—you ain't goin' nowhere." Or, as I'd put it: Have value, will travel!

However, favorable results can come even when dealing with a negative. Let's take a look at an example. You're talking to a friend. "I asked that girl from our advertising class out for pizza."

"What happened?"

"She said 'no.'"

Okay, in a sense, you lost—maybe not in a hurtful way, maybe not in a scarred-for-life-move-to-amountaintop way, but you lost and now it is time for community to kick in. You are not exactly a private person so you let everyone know. And your Facebook page is filled with you're-okay-so-get-on-with-life messages. At the same time, you are engaging your community with so-what-kind-of-girl-do-you-think-I-should-date conversations, and so on. It's a win/win, not the stuff of ol' William Shakespeare but, then again, life isn't Shakespeare.

Let us look at this a different way. She agrees to go out; the two of you enjoy the pizza and each other, and decided to go out again. A win/win, more conversation and fodder for the community! So where do you go from here? Something has to happen. Again, trite but true, nothing happens until someone sells something. In this case, you sell her on the idea of another date and then make the date happen. Organizations are individuals-write-large, with the same complexity of relationships and surrounding VRC of conversation and related interaction.

Savvy social media use builds community, and community is based on transactions, a never-ending stream of 'Wassup's" and replies. Transaction is one of the three components of community: *Interaction, Creative, and Transaction*. Strategic use of social media involves using all three to cement community. DailyCandy is a New York-based website that serves as the hub for emailed newsletters in a dozen cities for young female professionals who are looking for insider advice on restaurants, leisure activities, fashion, and sales. Its success (it was sold to Comcast for $125 million in 2008) is based on the breathless and chatty nature of its emails and the identification that its more than 2 million subscribers have for the site. The transactions begin with a 20-something signing up for the site and receiving access to its tips and sales (I give you my name and, in return, you give me information), a win/win for both sides. CrunchBase, a website compendium of technology industry information,describes it this way:

> DailyCandy, a free daily e-mail newsletter and website, is an insider's guide to what's hot, new, and undiscovered "from fashion and style to gadgets and travel. It is like getting an e-mail from your clever, unpredictable, and totally in-the-know best friend. The one who knows about secret beauty treatments, must-have jeans, hot new restaurants" and always shares the scoop.11

DailyCandy is one of the most successful and savvy users of database marketing today and its subscribers have formed a trendy and close-knit lifestyle community that has inspired many imitators.

BACK TO THE FUTURE, THE 87TH SEQUEL

How do we build a community and, once built, determine its efficacy for us? The first part we have, in fact, answered. Communities are built by ensuring that we involve the *right* individuals at the *right* time with the *right* conversation. Building a community works backwards from the target segment(s), which we covered in a previous chapter. Again, this is another remake of *Back to the Future*. We are continually going *Back to the Future* because so much of savvy social media strategy involves following the basics of effective marketing and results-oriented marketing communications. It's common sense, which, of course, is not all that common, especially not among our academic sorts.

When you are in the real world of competition and the dollars for your compensation have to be generated from the business of your organization (either commercial or non-profit) then social media quickly produce results if, and *only* if, you keep results in mind. Social media can be dramatically effective when we keep in mind the lessons of the past: results matter, you have to know where you are going in order to get there; a passionate affiliate gives birth to a passionate community, and a passionate community worked right always leads to *more.*

So we revisit some basic notions. Generally, what has worked successfully in marketing to affiliates in the past often, given updating for time and technology, works both in the present and in the future. If, after drilling down into your data, a certain type of target has reacted positively to the efforts of your organization in the past, that target is more likely to react positively in the future. Conversely, if you have chosen a target in the past that has not responded profitably to your appeals, then a different approach or target in the future is called for. This is how one vendor in the social space has put it:

> *In order to do this effectively for your business, you have to drill down into your activities and see where you've been getting the most bang for your buck. You need to determine what combination of tools, platforms and metrics are the best indicators of success when it comes to a return on your social media investment. Your efforts can pay great dividends, as long as you continue to measure and fine-tune what methods are producing the best results. If you conduct your social media marketing activities with results in mind, you'll outdistance yourself from the competition and reap the rewards that social media portends.12*

You have to find a way to measure the extent to which results matter so that you know you're getting results. One of the most difficult things to do is being able to account for what you've done. You'll be that much ahead of competitors, both personally and corporately, if you operate with an emphasis on measurable results.

PLAY IT AGAIN, SAM: RECENCY, FREQUENCY, MONETARY

Direct marketers introduced the notion of scoring the value of a customer along three dimensions: Recency, Frequency, and Monetary (RFM). We discussed this before, but it is worthwhile to emphasize this in relation to creating loyal customers. Loyal customers do more than just say nice things about your organization to others. They also show their loyalty in the form of increased affiliation and activity. For a consumer merchandising organization, for example, this means both buying more and increasingly more expensive products more often.

In the pre-digital days, direct marketers would use a formula that assigned scores for RFM. A "best" customer was one who had most recently, say, shopped in one of your stores, bought more products than the average of your customers, and whose average purchase was significantly above the others. In the digital age, we use a combination of actual transactions to measure the loyalty of a customer and include social media activity. A loyal customer is an active customer in all venues, which is why "the top priority of customer loyalty and retention programs is to increase customers spending, followed by reducing churn and turning customers into evangelists. Moreover... the correlation between customer experience and loyalty is high."[13]

We can now derive data from multiple sources (bricks-and-mortar, online E-commerce sites, responses to email offers, Twitter use, click-throughs, etc.) and track the various behaviors of individuals in the community. Put all of the interactions together and loyal customers are largely the best customers, our most frequent customers, and our most active customers. This works the same in non-profits: Loyalty should yield more donations more often in increasingly greater amounts. An affiliate is an affiliate is an affiliate, Philosophy 101. What works often makes the most sense. Or, as that great consumer merchandising philosopher Homer Simpson put it, "I can't believe it! Reading and writing actually paid off!?"

THE TAKE-AWAY

A savvy, which is to say, a well-planned and well-executed social media strategy focuses on increasing the number of loyal affiliates. Loyalty is not just one of those touch-feely emotional things; it is the core of social networks that bring growth. "Loyalty is as loyalty does," as Forest Gump might have said in the Academy Award-winning nineties movie of the same name, had he not said "stupid is as stupid does." Loyalty involves intensifying the emotional tie between a customer and a company, the savvy social organization has in place continuing strategies to intensify the relationships of its affiliates with the organization, and move as many of its best affiliates as possible into hyper-relationship.

Best customers buy more, best donors donate more, and best members use a fitness center more, take more courses, etc. Loyal customers are the best customers, and best customers are defined by the totality of their transactional activity. That's a significant part of effective social marketing, concentrating on moving affiliates into a loyal relationship that brings with it advocacy (which results in more customers) and transactions (sales, donations, event attendance, etc.). Affiliates are moved by focused and relevant social programs that include liberal use of direct and interactive marketing techniques to stimulate interaction and conversation.

Back to the Future. Who would have thought the DeLorean, the time machine built from a sports care that achieved cult status in the early eighties,would become a crucial weapon in the loyalty-building armory?

(ENDNOTES)

1   http://www.mycokerewards.com/howItWorks.do?WT.ac=mnuHIW_PO

2   ibid

3   http://digitalmedia.strategyeye.com/article/1u0Q4HxdJ3M/2012/03/22/spotify_fleshes_out_ecosystem_with_new_third-party_apps/1/

4   http://www.xomreviews.com/spotify.com

5   http://comicconmovie.com/

6   http://www.orangecounty.com/articles/says-23743-comic-mattel.html#

7   http://magazine.stevemadden.com/studly-mani-pedi-by-stephanie-stone/

8   http://blog.thismoment.com/2012/03/power-your-social-media-campaigns-with.html

9   http://www.engeljournal.com/define-big-idea-in-advertising/copywriting/2010/01/16/

10   http://www.adweek.com/news/advertising-branding/whats-big-idea-103274

11   http://www.crunchbase.com/company/dailycandy

12   http://sproutsocial.com/insights/2012/04/results-oriented-social-media/

13   http://www.1to1media.com/view.aspx?docid=33570&utm_source=1to1weekly&utm_medium=H&utm_campaign=04302012

CHAPTER 8

MONITORING & BENCHMARKING: IT'S ALL ABOUT THE DETAILS!

INTRODUCTION

SO, I CAN'T GET THIS SONG OUT OF MY HEAD:
Details, we love details,
What kinds of marketers love details?
Fat marketers, skinny marketers, marketers who climb on rocks
Tough marketers, sissy marketers, even marketers with chicken pox

Okay, so it is my version of the Armour Hot Dog song, which is about the nation's premiere picnic food and not about marketers. It makes the point that details are critical to an effective social marketing effort.

Here's the point: Those of us who are doing it digitally, who have done it digitally, who know what it is, to day-in and day-out, grow an organization and thereby opportunity…we respect details. We talk about details, all kinds of details, because we know that details provide both support for the formulation of winning social network strategies and the implementation of those strategies through tactical deployment of the appropriate tools.

So here's the lesson of this chapter: Because details are so relentlessly there, in every facet of our existence, we need to build on previous chapters and know where we're going before we have even a remote chance of deciding which details are relevant. Remember, the successful organization deals in a reality that rewards those who know where they are going and works back to front (from goals to where you are now) in creating strategies to get there. Those strategies are, at the lowest level, sets of relevant details to be managed and relentlessly inspected. When you are leading an organization, you like people who are able to handle details and separate the relevant from the irrelevant. Handling details separates the men from the boys and the women from the girls (or, if you're from a cat-loving market segment, the cats from the kittens) and allows strategic plans, once implemented, to succeed. Outside of the world of education and government, results matter. And those individuals who credibly promise to produce results add value to an organization and greatly help their careers!

Effective social communications programs, indeed, all programs that reach out to affiliates, are benchmarked and monitored. In the planning phase, savvy social media campaigns are linked to key activities, organizational responsibilities, and results daily, weekly, and/or monthly. This is an integral part of the preparation for the savvy use of social media, always asking (play it again, dear Reader!) "When is success, success?"

There's the savvy difference: Not everyone thinks in those terms. Instead, they wander, put in place an application here, a microsite there, and go on to the next tactic or strategy. But the savvy social user (you, dear Reader) knows better. You inspect, you monitor, and you follow the advice of that great scholar, Kenny Rogers: "You got to know when to hold 'em, know when to fold 'em," sang the sage whose last greatest hits album is the biggest seller the Cracker Barrel store gift shops have ever known. "Know when to walk away and know when to run." You just can't get this kind of wisdom in a classroom; but you can in this book, which makes it worth every dollar you've spent on it. After all, what is the price of wisdom? read on!

The bottom line comes down to knowing whether what you are doing is working and whether or not you will achieve the results you want. That's where measurement comes in, allowing you to compare the performance of your organization to planned performance or that of others. Too many people put a plan in motion, congratulate themselves on a job well done, and go on to the next project or assignment. But if you are reading this then you are, as the great sage Yogi Bear put it, "smarter than the average bear." You know how important it is to monitor what you're doing against your goals. The natural inclination is to get distracted; stuff happens, much of it unexpected, and that's why we always need to make adjustments. Social media, because of constant conversation, are ideal media for valuable feedback and real-time adjustment.

ADJUSTMENTS ARE NOT JUST FOR CHIROPRACTORS

Adjustments are not just for chiropractors. They are for all of us, applying to all activities big and small, because nothing happens *exactly* as planned and life often is one adjustment after another. The reason is (and I'm going to use my favorite word here) *stuff* happens all the time in both our personal and corporate lives. Stuff stands between you and achieving that perfect straight line to your goal. For example, I want to lose ten pounds in the six weeks before my Hilton Head vacation. After all, who wants to be mistaken for a beached whale at the shore? But I take my wife out to dinner and the dessert menu comes and, well, just this once. *Stuff* is a cheesecake, and cheesecake happens. In a few days, "oh, my favorite ESPN show is not on" and who can use a treadmill without ESPN? *Stuff* is no ESPN this time and *stuff* happens again…and again…and again, and I don't achieve my goal. Poor willpower? Perhaps, that and the vagaries of life, it happens. Life is a constant stream of *stuff* like flat tires, job losses, marketing miscalculations, operating slip-ups, changing markets, etc. Yes, there are awesome days and blessed events, but these are regularly punctuated by challenges and difficulties: *Stuff*. The savvy operator knows that you have to make adjustments. And just as the savvy operator makes adjustments, taking care to monitor the business of the organization, the savvy social media user keeps watch on the effectiveness of the social network of the organization. The savvy user monitors social operations and compares progress with affiliates to both the plan and the performance of other organizations using similar strategies and tools. When life hands the savvy operator a lemon (as evidenced by monitoring of results), he or she adjusts, making lemonade.

Here are some terms we need to know to play social chiropractor and make those adjustments:

Metrics: Metrics are numbers, percentages, and ratios used to count the relevant characteristics in all of the activities that, together, comprise an organization's marketing and operations plans. We measure differences with the numbers, giving us an idea of whether or not we are on target, reaching our goals. Let's say you are the manager of an organic food store that gets in a special shipment of bean sprouts grown in Iowa. This is unusual, because Iowa is not known for sprout farming and an Iowa entry in this category is an oddity. But these are special bean sprouts nourished by combining the flatulence (also known as farts, but only in a coarsened popular culture. Wink wink!) of Iowa cows with hydroponics. The sprouts are farmed in water that organically fertilizes the spouts while assisting in the global warming fight. As a savvy organic food store manager (yes, I know, bean sprouts do not have the same effect on brain cells as good ol' fashioned red meat, but you can *pretend* organic food store managers are savvy), you know this special shipment will send your targeted female shoppers into frenzies of delight from the bottoms of their Birkenstocks right through to the graying roots of their unkempt coifs. So you send a promotional email to your customers, offering a free bumper sticker ("My Other Car is a

Horse) with every purchase of the sprouts (which are priced like Wagyu beef especially for the granola set). Coming up with *metrics* for the campaign allows you to monitor your success. You may record how many emails you sent and track how many shoppers use the barcoded coupon to pick up the sprouts; or you may compare the number of comments with the keyword "sprout" or "bean" or, for that matter, "flatulence," all of which are important parts of the marketing strategy for this target.

Choose metrics carefully to reflect progress toward a goal. *Spoke* applications drive traffic; the ubiquitous click-through (click on a link that takes the user to a different page) count is a good beginning, as are other standard measurements too numerous to mention here. But it is worthwhile to consider not just the number of, say, responses to emails or re-tweets, but the amount of engagement as measured by buying, donating, or a similar core activity of an organization. Metrics should be relevant to your goal or objective. If you want to stimulate conversation, for example, then the number of comments on your blog would be a valuable metric, as would the number of re-tweets on Twitter with comments. Or, if you are a non-profit seeking donations before the tax year ends, you tweet the opportunity to contribute before the New Year, track the click-throughs and link donations, to *spoke* application activity, to monitor the ultimate effectiveness of your campaign.

The metrics you choose are determined by your goals and the various social media applications chosen for the online presence of the organization. However, do not fall into the more-more-more numbers trap: Social media, because every application allows a multitude of numbers to choose from, generate more numbers and therefore metric opportunities than an organization needs. Many of the measurements needed (and many that are not) are available with the click of a mouse and are called *web analytics*. Here's how Webopedia, the respected online reference source for the information technology trade, explains it:

> *Web analytics* is a generic term meaning *the study of the impact of a website on its users*. Ecommerce companies and other website publishers often use Web analytics software to measure such concrete details as how many people visited their site, how many of those visitors were unique visitors, how they came to the site (i.e., if they followed a link to get to the site or came there directly), what keywords they searched with on the site's search engine, how long they stayed on a given page or on the entire site and what links they clicked on and when they left the site. Web analytic software can also be used to monitor whether or not a site's pages are working properly. With this information, Web site administrators can determine which areas of the site are popular and which areas of the site do not get traffic. Web analytics provides these site administrators and publishers with data that can be used to streamline a website to create a better user experience.[1]

What's available? More than you'll ever need. Google Analytics alone can generate hundreds—thousands, with the premium service—of opportunities for comparison. The lesson: measurements and $4.85 will get you a venti Mocha Light Frappuccino at Starbucks; measurement alone will get you nothing. Choose carefully and stick to your knitting (another one of those literary metaphors, thereby making this book a bona fide educational endeavor).

Benchmarks and benchmarking: Never do for yourself what someone else can do better and, if someone is doing it better, figure out the why's and how's through a purposive comparison process, and then find ways to apply the smarts of another to your organization. As a veteran of media and consumer marketing operations,

I've learned to give others the opportunity to show their talents at marketing, strategy and operations, and leadership; I have rarely been disappointed. This also applies to knowing things; there are all sorts of people out there who know *stuff* and who can give us the benefit of their knowledge.

They almost always come up with something better than I would have developed by myself, a lesson in leadership. So, let's take a look at a foundational definition of benchmarking from businessdictionary.com and get a leg up on this topic. Benchmarking is,

> *A measurement of the quality of an organization's policies, products, programs, strategies, etc., and their comparison with standard measurements, or similar measurements of its peers. The objectives of benchmarking are (1) to determine what and where improvements are called for, (2) to analyze how other organizations achieve their high performance levels, and (3) to use this information to improve performance.2*

This is the active process of comparing where you want to be with where you are and, in our context, doing something about it, getting results. I can't repeat that last word enough: results, results, results, results, and more results. Organizations hire individuals in the belief that they, rather than others who they did not hire, can get results. Organizations, more often than not, promote individuals who produce results. Measurement, during all parts of the process of working a strategy, is designed as checking behavior for the organization, against the plan that was put in place, and the operations of similar companies attempting to achieve similar objectives in a category. The goal for the savvy social media user: results, world-class results, extraordinary results.

Albert Einstein is said to have defined insanity as "doing the same thing over and over again and expecting different results." He didn't—it's one of those urban myths.3 Butthe social media Einstein, mythical or otherwise,seeks continuous improvement in strategy and social operations, and incrementally better results. An organization benchmarks to find strategies and tactics that will produce more results at less cost. It seeks continuous improvement so as to get better, not just do the same thing over and over again. In the context of social media, we use benchmarking to achieve the following:

(1) See how effectively others in our business category are using a variety of social applications and strategies, at the same time comparing it to our approaches; and

(2) To compare how we plan something to work versus how it actually *is* working. Nothing works quite as planned (see life, marriage, dating, career, etc.). Organizations are simply individuals writ large, and sometimes the efforts of an organization are blessed, other times cursed. Corporate or individual, such is life.

For example, lettuce return to bean sprouts (Dad joke). The organic foods grocer we mentioned earlier has savings area, of a *hub* website, devoted to the many vegetables that are catnip to the Birkenstock set (mixing a metaphor to prove that this book is, indeed, wisdom on steroids). Digital coupons are being redeemed at a rate of one coupon for every 100 customers who click through the site; meanwhile, a competitor is getting a redemption rate of one for every 50 affiliates passing through their microsite, part of a set of sites in the competitor's hub. This is twice as good as our organic grocer. If the organic grocer is savvy, he or she will consider setting up a microsite for each category of vegetable and break out the coupons from the store website, assigning them to the appropriate microsites, just like the better-performing competition. The savvy operator knows how competitors are performing with the same approaches, goes "to school" on them, and changes approaches with the insights generated by comparing his or her organization to its competitors.

Benchmarking made easy! It is part of life: You wonder why you can't get anyone to go out with you when you notice that a classmate dates just about every weekend. So you observe and discover that he offers to pay for movies and pizza, whereas your approach, "You wouldn't wanna buy me a pizza later, would ya?" has consistently produced no takers. Suddenly it occurs to you that, in fact, you come off much like that famous French skunk, Pepé Le Pew, who said of himself, "A pitiful case, am I not?" So you change your approach, get better results, and you owe it all to benchmarking!

Benchmarking is not only used to compare the performance of one organization with others but also to compare planned performance with actual performance. Strategy demands assumptions as to performance, responses, all the activities and interactions that affiliates engage in as part of the community surrounding an organization. Sometimes, however, a plan for a promotion is based on a response rate of 20 percent and the actual response rate is closer to ten percent or a plan of 4.1 percent that actually achieves 1.3 percent. Why the difference? Incomplete knowledge on the part of individuals within the company when they put the plan together, perhaps; or market conditions not accounted for, such as a recession, are holding down responses; or those responsible for putting together the campaign engaged in a bit of wishful thinking to obtain internal support (my usual comment to those who reported to me was a cold smile accompanied by "It's not nice to fool Mother Nature"). There are many reasons why "actual" differs from "planned," but the savvy social marketing user checks his or her assumptions with reality, and then adjusts accordingly.

Here is another and more specific example: An organization uses a variety of applications to drive traffic to its hub website. When setting up an application to use as a *spoke*, an organization wants to know, above all, if it works: Does it drive traffic in the quality and amount planned? If Pinterest is expected to drive 20 percent of the traffic to the hub of an organization and delivers 10 percent, what happened? Was it how the organization implemented the plan, encouraged or did not encourage conversation, or the content of the "Pins" (the graphics 'pinned' to the virtual pinboards) in the application? Benchmarking compares planned to actual results and facilitates a process of "drilling down" (acquiring an understanding based on breaking down the various aspects of a tactic or tool in relation to a strategy and projected results) to find insights and solutions.

Engagement: This multi-step metric combines activity and effect and allows insights into the real story of the affiliate, his or her attraction and loyalty for the organization and its activities. Savvy social media programs produce more than click-throughs, page views, and a host of other popular digital measurements. All of these, for an organization to succeed strategically at social media, must lead to other activities that demonstrate, first, the attraction of the target for the organization followed by growing loyalty and, ultimately, advocacy. This is engagement.

Engagement metrics, which are rarely a single ratio or number, provide an assessment of the effect (the emotional tug) and activity ultimately produced by traffic counts. Engagement metrics tell us, in the immortal words of broadcast pioneer Paul Harvey, "the rest of the story." And it is the rest of the story, after the spokes drive the traffic, which is important. Avinash Kaushik, whose official title at Google is Digital Marketing Evangelist, put it this way:

> So what actually matters in Social Media? Not the number of Friends / Followers / Subscribers. Not the number of posts / tweets. Not the ridiculous Followers to Following

ratio. Not the... well there are so many horrible ones to choose from. What matters is everything that happens after you post / tweet / participate! Did you grab attention? Did you deliver delight? Did you cause people to want to share? Did you initiate a discussion? Did you cause people to take an action? Did your participation deliver economic value? The "so what?" matters!4

"The 'so what' matters!" Neither Einstein nor Homer Simpson could have said it any better. Engagement metrics tell a critical story, how connected the organization is to its affiliates and which tactics are most efficient at producing engagement. What percent of those coming through a particular type of content go on to buy, donate, or join? How many YouTube viewers later advocate the company's products to friends (as judged by combining YouTube views with response to, say, a "family & friends" promotion)? Measures that relate to engagement are rarely simple, necessitating a combination of metrics and a variety of linked activities, but they are well worth the time to construct. Every *spoke* activity and *hub* feature should contribute to growth of the core activities of the organization; that is, the affiliates of the organization buy more, sell more, bring in more activity-producing friends who go on to bond with the organization, etc.

Measuring engagement is difficult, but worth the effort. You have a certain number of readers of your blog, what percentage of those readers make comments? This is just the beginning. Now it is time to peek behind the curtain: those who comment tend to fall, perhaps, into the high-donation quartile, while other spokes (Instagram, for example, a photo-sharing application) produce affiliates who more often than not fall into the lower two donation quartiles. Most applications and sites have tools that provide organizations with detailed analyses of users, while many third-party tools are available for measurement. Obtaining metrics and producing statistics is the easy part; selecting the key metrics that combine to measure actual engagement is more difficult. Become familiar with the web development tools and applications that you use, select among the many tools and statistics available, and compare the metrics that result in your goals; are you making progress? Do the metrics point toward your best, most engaged affiliates?

However, take care that you don't succumb to a common disease for many users of social media once introduced to the multitude of easily available metrics: analysis paralysis. Analysis paralysis afflicts many users as they discover that the digital world is also a tracking world in which virtually any online activity can be counted, measured, and reported. Numbers, numbers everywhere and not one statistic that makes sense. Every metric you generate should measure the progress of the strategies of an organization toward a concrete set of goals, all resulting in growing loyalty and growing affinity on the part of affiliates. If this measurement chapter were a movie, it would star Jack Nicholson and be called *A Few Good Metrics*.

Count results, not the amount of metrics. Results count. While metrics are simply a way to ask the question, that former New York City Mayor Ed Koch achieved fame for asking of his constituents, "How am I doing?" Metrics can provide the answers that a savvy operator in a digital world needs.

And answers done right, produce results; and results done right produce growth.

BABY, BABY, CAN'T YOU HEAR MY HEARTBEAT?

But how do you know you're on the best course? How do you know that your savvy is showing? Metrics play a big part. But much depends upon what you measure; a critical part of the definition of savvy for us is "customer-centric." A customer-centric organization is focused on gaining an intimate knowledge of its affiliates. Listening through social media to customers offers the best path to success in a socially networked world. An organization that bases its metrics and benchmarking on constant evaluation of its customers through their conversation and interaction is an organization that positions itself for the kind of market understanding that propels growth.

Unilever's Dove Campaign for Real Beauty is an example of social marketing that has driven a company to extraordinary growth, dramatically outperforming the health and beauty category since 2004. Dove and its agencies monitor a set of metrics, that together, track engagement, which increased by 12 percent. Two years after the introduction of the campaign, share of market increased by more than 33 percent and Dove outperformed competitors on key media benchmarks by almost 80 percent.[5] Every year since, Dove has seen category-shaking results. The campaign is founded on customer insights, beginning with the original research that listened to women complain that media portrayals of women were unrealistic and damaging to their self-esteem. The social-based 24/7 Dove campaign has integrated websites, microsites, and a legion of other social media applications, with Unilever product managers relying on a continual stream of metrics that measure return on social investment. Here are some examples of Dove monitoring: the number of women signing up for webinars, traffic on its various sites, YouTube views, viral videos, webinars, online hubs for activity and coordination, etc.,

This is what benchmarking and the use of metrics in the social media world is about: Keeping customers under a metric microscope that allows managers to adjust strategy and tactics in real time rather than the traditional and oh-so-academic approach of sitting down at the end of the campaign and assessing, which is educator-speak for "wake me up when the action is over and I'll offer suggestions for doing it differently and, by the way, don't confuse me with all that real-world stuff!" Savvy social media use requires organizations to measure and understand affiliate activity, in such a way as, to be in tune with the very heartbeat of the customer at the time of the beat, and not "assess" when the patient is dead. The British band Herman's Hermits climbed to sixties fame with a catchy song that asked, "Baby, baby, can't you hear my heartbeat?" Successful social marketing answers that question with a resounding, "Yes, Mr. or Ms. Customer, we can hear your heartbeat, using a series of metrics that tell us what you think, where you're going with our organization, and how we can get closer and engage with you." Admittedly not as lyrical as Herman's Hermits, but listening to the voice of the customer provides continuous and actionable information. Baby, baby, can't you read my metrics?

Affiliates (customers) are the lifeblood of an organization. Famed management consultant Peter Drucker put our priorities this way: "The purpose of a business is to create and keep a customer." Listen and act, test and react, base every move on the customer. This is possible now that the Social Media category allows an organization to listen and interact with customers 24/7. Metric and benchmarking programs can provide an understanding of how current strategies work and how future strategies need to be constructed by concentrating on measures that go to the heart of consumer activity and allow organizations to understand what affiliates are saying through behaviors and accompanying conversation. It is important to interpret it in the right way, as emphasized by the experts at Peppers and Rogers Group:

> *"Customers are implicitly and explicitly telling organizations what they want, in different ways," says Wilson Raj, global product marketing manager for customer intelligence*

at SAS [the official name of the software engineering giant that once was referred to as 'Software Analysis System']. Yet, although customers talk a lot about brands, this dialogue is not always directed toward them. This is sometimes because customers don't want to tell organizations what they think, or because they don't know how to best put this information across. According to Sandra Zoratti, vice president of global marketing at Ricoh, some customers make overt or covert omissions when it comes to telling companies what they think. This means that organizations have to be alert to the nuances in customer behaviors and interactions to find out what customers aren't telling them, as well as how they can apply this knowledge to improve their business.6

Monitoring metrics and benchmarking provide a living portrait of an affiliate in relation to the organization and its activities. Organizations can now observe and listen to the customer in real time as he or she relates to the organization. However, it is important that an organization listens and acts on the feedback coming from social media activities; the extraordinary power of the latter is due, in large part, to the ability to listen and act on customer activities and attitudes as they happen, making adjustments is necessary. A significant part of the fall of the newspaper industry has been its refusal to listen to affiliates or readers. As head of marketing and sales for a media company, I worked for years with newspaper staffs that were determined not to listen to customers. The attitude of an editor, at our Madison, Wisconsin newspaper, was symptomatic of the problems of the industry as he refused to even look at the results of a reader survey, dismissing it with a blunt, "Who cares? My job is to tell them what they need to know, regardless of what they think." In other words, shut up and read, dear customer! The consequences of that attitude are now apparent to all. Google "newspapers dying" and you get 24.6 million hits, compare that with "disco dying" and you get only 10.9 million hits. Why the comparison? So you might say that consumers are about 2.5 times more likely to take up disco than they are to read a newspaper in this, the digital age (or maybe not…).

Now the digital age has unleashed an army of competitors and, even in a town dominated by the I'm-too-brainy-for-my-Prius denizens of the University of Wisconsin, savvy organizations embrace what has come to be known as "Voice of Customer" (VoC) programs. Peppers and Rodgers Group (as you can see, I'm a fan of their results-count approach) notes:

[A]n increasing number of companies are taking action by fostering customer-centric cultures that enhance satisfaction and drive bottom-line profitability. A cornerstone of their approach centers around gathering, analyzing, and acting on the voice of the customer (VoC). By carefully crafting VoC programs, and leveraging customer feedback tools to support it, companies are increasing loyalty, lowering service costs, and boosting profits when and where it matters most.7

Conversations are manifestations of customer voice, providing organizations with the feedback necessary to construct effective multi-media approaches with social media programs at the center. In fact, the current notion of integrated marketing communications (IMC) is, to a large extent, passé: The digital social revolution has put online conversations at the center of all coordinated marketing communication. Social media strategies and tools are no longer one of many ways to reach the affiliates, of an organization, as the tradition-bound definition assumes with IMC "designed to make all aspects of marketing communication such as advertising, sales promotion, public relations, and direct marketing work together as a unified force, rather than permitting each to work in isolation."[8]

Rather, the facts on the ground tell us that the online socially mediated world is the central organizing force now being used by organizations large and small, commercial and non-profit, business-to-business and business-to-consumer. I could go on and on (I am, after all, now a professor) but the single can't-be-denied, no-doubt-about-it, so-obvious-even-a-professor-would-get-it conclusion that comes out of even a cursory glance at the real world around us is that every organization must have an online social presence as its organizing force for the world around it. If you remember, we began the book with a variation on the great Homer Simpson quote on donuts: "Donuts. Is there anything they can't do!" We changed it to "Social media. Is there anything they can't do!" Well, we're in the tenth chapter and the answer is still "Not that I can see!"

> *Social media, done right and through conversation, offer a continuing and deep understanding of the relationship of a customer/donor/member to an organization and are the single greatest source of research for the strategies and operations of an organization. Social media provide an unparalleled means of organizing affiliates around the activities of an organization, with conversations offering opportunities for an organization to engage and guide its target segments. All other media serve to reinforce and support this central organizing force for the company. Although traditional media can channel great gulps of a market to the organization, it is social media that engage and bind the target far beyond the short and traditional frameworks of Old Media strategies. The Social Media category does this through the power of engaged and purposive digital, interactive conversation, with monitoring metrics and creative strategy included in the package.*

When all is said and done, Integrated Marketing Communications, IMC is so, so, *yesterday*. Today and tomorrow: *Integrated* Conversational *Communications, ICC. Integrated Conversational Communications, now that's got legs!*

ZEN AND THE ART OF CONVERSATION MAINTENANCE

Bet ya think I don't do anything but sit around and watch television and movies! Well, I read a bit, too (usually on my smartphone during faculty meetings), a remnant of a lifelong love of fine literature: comic books. I did once work my way through *Zen and the Art of Motorcycle Maintenance*, the Robert M. Pirsig account of a cross-country motorcycle trip punctuated by philosophical discussions, many centered on Plato (a philosopher who, I think, may have been related to Kato, the Green Hornet's assistant). It was the only book I read in a required philosophy course during my undergraduate years. Although there were seven other books required. But you're talking here with someone who, when asked by a rhetoric professor if I understood the implications of Plato on communications, replied "Absolutely, Play-Doh revolutionized the way we relate to the world!"

What does this have to do with conversation, social media, and organizations? Everything! Conversations have become the marketing lifeblood of growing organizations. Savvy socially mediated organizations and affiliates become a single entity through conversation, hovering together within a virtual cloud in which branding and positioning happen in real time. And just as Pirsig embraced a view of a world in which truth arrived as experience mediated by bursts of creative thought and insight so too must the savvy social media user embrace the social net as a place where conversation provides organizations with continuous "bursts" of "truth" about the affiliate/organization relationship and momentary eruptions of creative approaches to align organizations and affiliates. These "truths," which arrive in the form of insight and understanding from the cloud of relationships offered through conversation, then fuel the communications strategies of the organization. Some of the strategies involve Old Media such as billboards, television, direct mail and other means of episodic and finite promotions. But *all* of the strategies are implemented within the greater context of continuing social conversations. These conversations supply the *stuff* (that word again) of savvy communications, as all social media applications and tools provide the capacity for creative strategy, the ability to measure the truth of results, and the means to effectively adjust strategies.

The social application industry daily introduces new approaches to measuring the effectiveness of conversations. A continuous stream of new social applications, enhancement of existing software, and an emphasis on creating new and enhancing existing measurement toolboxes for organizations to monitor return on digital media investment, provides organizations with the means to make conversation the core of marketing strategy. It is what they do, how they work, a continual creative process by developers that makes the applications more relevant and useful just about every time you turn around. Here's a past announcement from Facebook, which is typical of the Larry the Cable Guy "git'r-done worldview" of the Social Media category:

> *Facebook has introduced several new metrics for advertisers based on the actions consumers' take after seeing a Facebook ad. Previously, the social network didn't provide insight into what consumers did after clicking on an ad, making it difficult for you to see the impact of your advertising campaign. Through the old system, you could only see data about the number of Likes a Page received as a result of an ad. The new metrics can be found in the ad dashboard called Actions, and will include comments, shares, app usage, and Credits spent . . . By giving you more detailed data, Facebook hopes you will move away from establishing Likes as a goal and is encouraging more focus on engagement. The update doesn't affect the pricing model; Facebook ads will still be sold on a cost-per-click or cost-per-impression basis. Additional insights will help you measure the success of specific marketing objectives, as you're now able to distinguish which actions came organically or through paid media.[9]*

Time to wax poetic: This is great poetry in action, a *Lovesong of J. Alfred Prufrock* (the classic T.S. Eliot poem that most of us are assigned to read at some point in our school years but few actually read) for the virtually empowered. Call it *Lovesong of J. Digital Media*:

> *Good, better, best.*
> *Never let it rest.*
> *Until your good is better*
> *And your better is best*

Take that, T.S. Every second, every minute someone is trying to better refine an app, introduce a more involving technology, turning an iPhone 4, into an iPhone 4S, into an iPhone 5. Now multiply this effort by thousands, perhaps tens of thousands, as social software developers and managers continually work on ways to integrate conversation into the operational and marketing process, and you understand that customer-centric strategies are no longer limited by tools, but only by imagination. What metrics are useful? It depends upon the social media you use, the engagement you seek, and the needs you have for inspection. However, keep in mind that a return on your social media investment means growth and, consequently, expanded opportunities.

The savvy social media organization selects metrics in the same way it selects applications and development tools: Use what works. Again, work backwards from the desired results, pulling out the tools and metrics needed to "git'r done." Wherever you turn in the world of social media, the emphasis is *results* by sharing and hyper-relationship. From the growth-pushing developers at Foursquare, the mobile app through which individuals interact with their environment and share the experience put it for businesses:

> Millions of times a day, people use foursquare to check in and share where they are. Whether checking out a new restaurant, meeting up with friends, or visiting a favorite boutique, they are chronicling and sharing their adventures. As a business or brand on foursquare, you can be a big part of that experience.[10]

Again, this could be the message from any social application developed at any time: We love growth, we love organizations to market and operate in such a way as to drive growth and the affiliate experience, and we love getting stuff done! Period. The end.

THE TAKE-AWAY

Trite but true: You get what you inspect, not what you expect. The savvy social media manager puts digital conversations with customers at the center of his or her growth strategy. This increases the manager's ability to monitor programs and make appropriate adjustments. A key strength of the Social Media category is the ability of its applications to track activity and engagement, allowing an organization to determine in real time whether strategies and tactics are having the desired effect.

Metrics 'R Us, that's the retail chain that social media would open if social media could open a retail chain (a sub-grouping of the Woodchuck category, in which we ask "How much wood could a woodchuck chuck if a woodchuck could chuck wood?"). Monitoring and benchmarking capabilities are part of the fabric of Social Media, which makes the category dramatically more productive for an organization to use as a foundation for its operations and marketing capabilities. Wrong turn? Okay, take a left at the light, go three blocks, turn at the Old Media dumpster and you're back on track. Don't like the way customer conversations are shaping up? What do the metrics tell us? You get the picture, I can *feel* it. It's a social thing.

(ENDNOTES)

1   http://www.webopedia.com/TERM/W/Web_analytics.html

2   http://www.businessdictionary.com/definition/benchmarking.html#ixzz1ubQF3DH7

3   http://nymag.com/news/intelligencer/insanity-quote-2011-9/

4   http://www.kaushik.net/avinash/best-social-media-metrics-conversation-amplification-applause-economic-value/

5   http://thearf-org-aux-assets.s3.amazonaws.com/ogilvy/cs/Ogilvy-09-CS-Dove.pdf

6   http://www.1to1media.com/view.aspx?DocID=33464&PreviewMode=full&utm_source=1to1weekly&utm_medium=H&utm_campaign=02272012

7   "Don't Be in the 4%" White Paper, Peppers and Rodgers Group, 2012 (accessed May 2012, http://www.1to1media.com/downloads/vovici_prg_voc%20white%20paper_FINAL_021712.pdf?%E2%80%9D

8   http://marketing.about.com/od/marketingglossary/g/imcdef.htm

9   http://sproutsocial.com/insights/2012/04/facebook-advertising-metrics/

10   https://foursquare.com/business/

CHAPTER 9

APPLICATION POLICY: TEACHER, MOTHER, SECRET LOVER

INTRODUCTION

AND SO WE NEAR THE END...AND BEGINNING. HOW'S THAT FOR MELODRAMA?

The sage of Springfield, Homer Simpson, once said of television," Television! Teacher, mother, secret lover!" A savvy organization can say the same thing about social media: Teacher, mother, secret lover. Teacher: The social network surrounding an organization is the most powerful tool that has ever existed to learn in real-time what customers are thinking, feeling, and doing, while at the same time, schooling affiliates in the values, products, and services of the organization. Mother: Through a digital social network, the organization has the ability to let its affiliates know it cares, that it wants its customers to succeed, and that the individuals conversationally connected to the organization are first on its mind, a healthy and nurturing maternal relationship. And secret lover: "You are loved" is the background message of all savvy communication to the affiliate, an undercurrent in all conversations, and the foundation of all moments of truth between the customer and the organization.

Social media approaches are extraordinarily powerful in effect, but notoriously difficult to coordinate and control. It is one thing to manage that which you actually have control over, like billboards and television spots; you control the paid media and advertising content. But to put it bluntly, this chapter focuses on the out-of-control challenges of social media marketing. We are asked by a digital environment to shape the conversation and activities of affiliates, who, theoretically, are free to choose the actions and reactions that comprise a company's marketing! Sure, says the customer, you can bring up a topic for conversation—but maybe I'll ignore the suggestions of your organization or misunderstand your point—so talk away and let's see what happens!

This chapter focuses on applying the strategic and tactical approaches of previous chapters, and integrating the marketing worldview of the savvy student or employee with real-life applications (technology) in a marketplace where affiliates don't have to listen. Digital technology is feature-rich and social media are ubiquitous. We offer examples that point the way not only to the use of existing tools but also demonstrating how other and future social media applications can be integrated into existing interactive strategies to maximize the probability of success. We can't guarantee success; failure is simply part of the human condition and, as another great thinker, former Yankee catcher and manager, Yogi Berra (not to be confused with his cousin, Yogi *Bear*) has pointed out, "If the world were perfect, it wouldn't be." As I've said so many times: You can't know everything, but you *can* learn how to think, how to approach problems with a worldview that is immensely practical and extraordinarily effective. By doing so, more often than not, you will succeed in a marketplace where results count.

This chapter relies on the groundbreaking theoretical work of Homer Simpson in this field—Teacher, Mother, Secret Lover—to summarize our savvy approach to social media. In this, an organization deliberately and successfully shapes the thoughts, attitudes, and behavior of the affiliates who, in the age of digital interactive media, are vital to its existence. Social Media provides us with the ability to not just pay attention to customers, but to know them intimately and provide the kind of love and mentoring ordinarily found in the best of families, and that is now seminal to the organization/affiliate relationship.

So many people lament the power and ubiquity of social media and digital technology. "You can't get away from it," weeps the university professor. "I can't get my students to put away their smartphones," says another. "It's taking over promotion," whines the marketing communications manager. "It's standing in the way of wholesome human relationship and ought to be banned," says the Ivy League deep-thinker. "It is forever changing our culture," whimpers the seriously deep-thinking guy glued to the seat in the corner of the local Starbucks.

Meanwhile, Homer Simpson stands squarely atop the world of social media, shouting loudly and powerfully so that all can hear and take the first steps toward savvy social marketing, "D'oh!!!" Homer nods.

D'OH!!!

"No kidding."

"How right you are."

"Hey, this is how it is, so get over it!"

These quotes sum up the reality of the role of social media in driving corporate growth and are encapsulated in the simple Homer-ish expression: "D'oh!" "D'oh" is fraught with meaning, laden with significance. It is the expression of support for an approach that has become so powerful and common, has so many features and benefits for the organization operating in a digital world that the savvy operator makes social principles and conversation integral parts of corporate growth strategies. "D'oh!" represents a way of looking at the world that, in essence, says: This is reality, the power of social media. Conversations are the primary means by which individuals have communicated in the past. Now, thanks to the digital world brought about by human ingenuity and relatively free markets, organizations may join with humanity in using conversation to establish mutually beneficial relationships with individual customers, donors, and members. So deal with it!

And grow. Organizations use growth to acquire more resources, to do ever more and greater things. After all, it is entirely human to want more from the world around us and organizations are merely individuals writ large. And *more*—more love, more intimacy, more responsibility, more customers, more resources—brings rewards for greater amounts of people who, in turn, may do more for those around them, who do more…you get the picture! Growth done right, like social media done right, offers the satisfaction of social and economic opportunity. Organizational transformation consultant George Land, in his groundbreaking book aptly titled *Grow or Die: The Unifying Principal of Transformation*, explained that at "the root of the principle of transformation lies a single concept: growth—the most basic and universal of drives through which . . . all biological, physical, chemical, psychological, and cultural processes are intrinsically equivalent."[1] We crave more, we want more, we need more, and so we search with a built-in restlessness for connection to others in all parts of our lives, both personal and corporate.

Forget the academic theories, the mutterings about peace and quiet, and the vague longings for untethering from the social net that are trendy in some circles. One of the most striking characteristics of much of the academic world I have entered is its disdain for the commercial world that I had inhabited for most of my adult life. The comments and rants are comic for their echoes of the Pink Floyd-ish wisdom once floating about the smoky haze of the Depraved State University dorm rooms in the seventies. But note that even those who claim otherwise (stop advertising, ban cell phones, why can't we just all get along and eat organic vegetables grown with unicorn manure. Waaaaah! Waaaaah!) desire connection, significance, and conversation.

The need for connection is universal and the savvy social organization understands this, integrating digital technologies into its marketing and operations. You, dear Reader, may go out into the world and use its 24/7 window on the soul of the customer reach out to your advantage and that of your organization, increasing revenue. Or you, dear Reader, are already in the world and will use social strategies to grow your company, hitch your wagon to growth-inducing conversation, and garner the kind of praise and reputation for savvy that rewards you with increased responsibility and compensation.

This stuff works, this point of view works! Based on customer feedback, you can create strategies that take advantage of your customer-centric understanding of the individuals affiliated with an organization. They want a closer relationship, and their conversation is a personal manifestation of that need for closeness as it relates to your company. And because of that need for closeness, the organization is capable, through socially mediated strategies, of meaningfully bonding with affiliates over the products, services, perceptions, and impressions surrounding each particular organization.

So what does this mean? Generally, "growth done right" offers greater significance for greater amounts of individuals. DailyCandy.com is one of thousands of examples of organizations building on an intimate, socially mediated understanding of its affiliates. The trendy style website for women has grown dramatically over the beginning of the twenty-first century from a single woman on her computer in a small Manhattan apartment to an enterprise valued at more than $123 million with hundreds of employees, all of whom share in the passion of their affiliates and are not afraid to get both mushy and gushy in content, in conversation, in growing compatibility and engagement.

As the organization has grown it has added more *spokes*, driving traffic and pushing conversation through Tumblr, Twitter, Facebook, and YouTube, just to name a few of the applications that lead to its powerful and refreshingly authentic *hub* website. You can see and feel the love reflected on its website and throughout the *spokes*. The enthusiasm of young professional women is on display online in text and graphics and video. Why? "Because at DailyCandy, our editors relentlessly seek the genuine, the unique, and the next. We love the thrill of the 'find.' We never stop seeking what's new in fashion, food, and fun. Discover what you love at DailyCandy, and share your finds with us."[2] "Share," "new," "love," "genuine," and "unique" are not just words but genuine expressions of hyper-relationship that partner with technology and an understanding of affiliates to grow the organization and opportunity for both affiliate and organization.

Savvy social media use begins with a commitment to growth, in customers, in affiliate-generated content, in inspection routines and metrics, and in training of individuals within the organization to conform to the worldview necessary for effective socially networked communication. It calls for employees who are as emotionally connected to the organization as the affiliates of the organization, and who can operate in an environment of relentless feedback, innovation, and emotional involvement.

Insight for savvy social users: Caring starts inside and works its way out. We can say of employees, what Casey Stengel said of his un-remarkable New York Mets baseball team on the way to astounding the baseball world by winning a division title back when major league baseball was America's pastime, "You gotta believe."

D'oh.

NO KIDDING!

There is a sense of the obvious in what we're discussing. The more a person talks with you, the more interested he or she is to you and vice versa, the closer you become. No kidding! That is covered by this bit of wisdom: If it looks like a duck, quacks like a duck, and walks like a duck, then it probably *is* a duck. Social media is ubiquitous because, well, it is everywhere. It is the primary means by which individuals connect with each other because just about everyone communicates online, through a combination of computers and mobile devices. And conversation is what we *do*. As another great philosopher, New England Patriot football coach Bill Belichik so wisely put it: "It is what it is."

No kidding. D'oh, and double-d'oh.

Savvy use of social media begins with the understanding that conversation counts, and that an organization must have a disciplined, coherent approach to messaging if it is to have an impact in a category that is always "on" and what people do naturally: converse. It relies upon a new business model necessitated by the digital conversation age. This model requires organizations, as one expert notes, to be "working it out in real-time, crowd-sourcing, transparency, and forward momentum [that] is both breathtaking and confounding, depending on one's capacity for stress."[3] Yes, an organization may still use Old Media (television and radio spots, out-of-home advertising, magazine ads) but these promotional approaches are increasingly being used as support and may even be considered simply another set of *spokes*. *Hub-and-spoke*, you remember, is what we suggested as the structure for a social media marketing capability. The *hub* can be a single website or social media application, or a set of websites, including microsites, and applications that involve visitors in the activities of the business; the *spokes* are the various sites and applications that serve primarily to channel and drive traffic to the hub.

In this scenario, a billboard along the highway drives traffic to an apparel retailer's website. There, a savvy social media specialist has used development tools that allow affiliates to download coupons, comment on the latest fashions, discuss the service at the stores, watch videos of a fashion show, and talk with other affiliates about the need to accessorize with ostrich plumes. A variety of applications and tools provide a reassuring and seamless opportunity for the affiliate to buy products from the website or find out where to shop at a bricks-and-mortar store. Meanwhile, many of the conversations include individuals who represent the organization, because a savvy social marketer understands that directed conversations are critical to effective social strategy, and individuals in the organization must participate in the conversation. However, employees of the organization understand, through training and the internal operating environment deliberately created by the organization, that their goal is to grow the organization through an increasingly engaged relationship with affiliates.

An engagement approach requires a willingness to spend the time and energy to deliberately shape the conversations that are taking place. Individuals within the organization must have guidelines for their participation in the continuous conversations surrounding the organization; policies that cover not only what is to be said, but also how it is said and what values and worldview are to be emphasized by the company. Greater connection requires a sincerity and visibility on the part of the individuals within an organization that has to organize, categorize, withstand, and understand thousands, if not millions of conversations, comments, and moments of truth.

What does this mean? The savvy social media specialist needs to be more methodical, disciplined, and purpose-driven than has been the case in traditional marketing communications. In addition, there must be processes in place that allow the organization to monitor and measure its progress in pursuing its growth goals. These should include a mix of hard measures (revenue, donations, click-throughs, contacts, etc.) and "softened" measures. "Softened" or soft measures typically combine hard and soft metrics (soft: measures attitude, satisfaction, happiness) to judge engagement, loyalty, and advocacy, and are usually measurements constructed of several metrics.

FAQ 14: Uh, oh—seems like some of this is vague. How are we to know if you don't tell us exactly what to use, what to do, and when?

"Welcome to the real world," said Morpheus to Neo upon the latter's arrival in "The Matrix." And so I'm saying this to you, dear Reader—welcome to the social media world, the real world where (like the Matrix) things are not always what they appear to be. That's the case with this whole business of knowing, as I said earlier, when success is success. In the good ol' days, when Old Media reigned supreme, success is what someone from the agency or media sales department said it was. We delivered the impressions (exposures to your ad) we said we would; this television program had the number of 25-54 year-olds you asked for; the billboard had 104,000 cars passing by...yada yada yada. The major problem with this is that these numbers were guesses based on samplings taken at various times and then turned into statistically constructed numbers for the benefit of the client organization. A more or less educated guess generated over three-hour "Mad Men-esque" lunches, but still a guess. This explains why a marketing communications campaign could be termed a success while the client organization didn't see increased demand for its products, services, donations, etc. The promotions worked but results were weak or nonexistent; the operation was a success but the patient died. In social media, however, we know when success is success: we've increased conversations by a certain percentage; we're getting more traffic to our site; more young women are buying between the hours of 1 a.m. and 5 a.m. on our website; we saw increased mobile "check-in's" at 75 percent of our factory store outlets, etc. You get the point: You need to define success based upon the actual results from the applications you are using to drive traffic through your spokes; the activities defined by the interactions in your hub; and the engagement you know will bring greater growth in the future. The answer is out there (Neo), dear Reader, and it's looking for you, and it will find you if you want it to.

HOW RIGHT YOU ARE!

The customer is always right: Trite but true, in a way. In other words, it is true at a deeper level. Yes, as anyone who has dealt with people in a service capacity (customer service counter lines, retail stores, church membership coordinators, "giving" executives, etc.) can tell you, customers are sometimes wrong, even obnoxiously wrong. A customer is wrong, for example, when she returns a pair of rode-hard-and-put-away-wet shoes eighteen months after purchase and demands a new pair ("these just aren't that comfortable anymore"). In this, the customer is unreasonable and demanding, has an inflated sense of entitlement, and is wrong, wrong, wrong. But the attitude on the part of the employee in a social era should be that the customer is figuratively right even when literally wrong.

The organization, then, must find a way to satisfy the customer, up to and including, offering a new pair of shoes or a refund. This acknowledges that a customer, an affiliate of an organization, is not just a stand-alone customer; rather, she or he is a part of a greater social network surrounding the organization, one in which the affiliate is, more or less, in communication with others quite a bit of the time, and the customer experience is now the stuff of a spreading network of socially mediated conversation. This means that a woman going home with a new pair of shoes offered with smiles and good cheer (even if undeserved) will be more likely to go online and let other affiliates of the company know about her pleasurable experience; to evangelize the organization and its activities to friends and family; use the event as material for future conversation. In other words, she has greater value for the organization having been loved than having been insulted (however well-deserved that insult!). The chances are that as querulous and difficult as she is, keeping her as a customer will accrue greater lifetime value to the organization, especially when the resulting conversations based upon the return and satisfaction is factored into the results. Will some take advantage of this customer-driven stance? Yes, but again, we're playing the odds, increasing the probability of more positive digital feedback and experiential results.

Social media strategies rely upon individuals within the organization using social media tools in an appropriate and positive manner to elevate the good narratives surrounding the organization and, to the extent possible, defuse damaging narratives. This depends, however, upon individuals within the organization understanding that, when they provide unsatisfactory service, they are not just insulting an individual; rather, they are insulting all of the individuals who surround that person in community, those who read his or her Facebook or Twitter posts or views Instagram photographs, or who see blistering reviews on Yelp and Urbanspoon, who let Foursquare members know an establishment wasn't welcoming, and so on through the continually changing lineup of social media at the fingertips of someone with praise or condemnation.

An example mentioned previously: A protest song, "United Breaks Guitars," went viral on YouTube after its author was given the runaround for a year by the airline's customer service division after his guitar was broken by careless baggage handlers. The song also became an iTunes hit; forcing United Airlines to recognize that the days of aggrieved customers suffering in silence were over. They then established internal social policies and approaches, including the monitoring of and communicating through social media.

The musician went on to write a book about the controversy appropriately titled *United Breaks Guitars: The Power of One Voice in the Age of Social Media*.[4] He advises people "that they have an ability to make change happen if they don't take no for an answer and can harness the power of new technologies at their disposal." The lesson is "that we're all connected . . . It means there's something there that makes it possible for us to connect with people if we just try."[5] Yes, individuals now have the ability to make change happen, as does the organization through the individuals that represent it. Anyone can have a conversation (even musicians), but not everyone can have the strategically correct and engaging conversation.

TEACHER, MOTHER, SECRET LOVER

Meanwhile, back at the ranch…

Enter the sage of Springfield, Homer Simpson, stage right (stage *correct*?).

Homer makes sense. Savvy social media use, when applied in ways that helps you personally earn the reputation as an employee that gets things done, assists an organization in getting things done (file under "Meet-or-Beat Our Goals" and "Wow, Did You See the Size of that Bonus!"). This requires a complex blending of social relationships with marketing and operational choices in pursuit of the growth goals of the organization. There are so many technology choices, so many applications, so many changes, and so many approaches to conversation, what's a savvy person to do? Quite simple for now: Keep reading.

That's why it makes sense to look at savvy social media marketing in the same way Homer Simpson looks at television: "Teacher, mother, secret lover!" The savvy social organization is populated by managers who are *teachers*. Teaching at its best helps students to successfully navigate the treacherous waters of life. The savvy strategist sets up a social network in such a way as to teach the organization how best to engage its customers and grow the organization while guiding (teaching!) affiliates in the most satisfying path to relationship with the company.

So teaching begins with organizing the marketing effort into a *hub-and-spoke* configuration. The critical activities of the organization that make it unique are represented in the hub. Again, the hub can be a website, a social application such as Facebook or Pinterest, or a central combination of these across which the activities of the company are spread. For example, a consumer merchandiser may have a corporate site used to inform investors while wholesaling merchandise; an E-Commerce site for retail and gift orders; or a microsite for loyalty club members. Or, perhaps, Facebook can be joined with an organizational website as two-site hub in the way the Sarah Palin organization has set up its Facebook pages[6] for political statements and its SarahPAC for donations and organizational news. [7]

Meanwhile, other promotional activities are used as spokes, with each spoke driving traffic in unique ways. Twitter, Reddit, Foursquare, and a host of others may be used as spokes to drive traffic to the hub, at the same time providing venues for continuing conversation. Spokes are generally social, but traditional promotional and marketing communications may also be used. Traditional package design can be used as a spoke by adding a reference to the company website to the package, and sending bricks-and-mortar buyers online where the website is designed to involve them socially in conversations about the brand. The same can be done for television and radio spots, magazine ads, even promotional items; all are the *stuff* of conversations and conversions and, although not social in nature, can be used as social venues for driving traffic and the content of conversation.

The savvy social user teaches, shaping skills while imparting knowledge relevant to the work of the organization and the hyper-relationship of affiliates, both for the organization and its affiliates. The organization must learn how to create and maintain a social media capability while customers must understand the value and benefits of the products and services of the organization. Bottom line: Organizations must be coached into effective use of social media as the foundation of their growth strategies, at the same time engaging affiliates or customers through a variety of technologies.

So, which applications, features, and software? It depends. How do you choose? It depends. First know where you are going and then decide how you'll achieve your goals. All of this is achieved by making the organization an integral part of the life of the affiliate and vice versa. Teaching a customer about the advantages of the products and services offered by an organization and then, based on that understanding, encouraging conversation. Who was it that said a social marketer's work is never done? Dunno. It's not easy but well worth the effort.

In the traditional and limited world of Old Media, you can generally control the content of your communications. But user-generated content depends upon customers who are not just taught about the product, but nurtured into relative and positive conversation. In savvy social media use, affiliates must be mothered into sharing the relevant worldview of the organization. As a good mother helps to shape the worldview of her children, the savvy social media organization shapes the conversation of its affiliates, spurring conversational growth and a shared affinity between the organization and its affiliates. It does this through creative use of social techniques, as illustrated by this successful Heinz campaign that was pointed to as one of the ten best social media campaigns of its year by an Advertising Age columnist:

> *Heinz partnered with the We Are Social agency for a U.K. campaign launched at the start of cold and flu season. It enabled Facebook fans to send sick friends cans of cream of tomato or chicken soup. For a $3 fee, a user could buy the can of soup, which was inscribed with the recipient's name and shipped, arriving at its destination within three to four working days. It would bear the greeting "Get Well Soon" and the sender's name.8*

Mother: Heinz provided a way for its affiliates to gently and positively engage with others around them. It did not leave it up to each affiliate to come up with an approach to cheering up a sick friend; rather, it provided the means through social media to do so in an approach that fit with the Campbell's Soup worldview. Social media tools may be used to shape the responses of affiliates, first teaching and then nurturing them through a process of learning the proper way of relationship.

Here is another example from a fashion magazine seeking to create and shape promotional conversation among its readers:

> *Belgian fashion magazine Flair built a Facebook app, Fashion Tag, to help users find out where their friends discovered their best outfits. Created by Duval Guillaume, Brussels, the app lets users tag friends' clothing and accessories and then post on their walls to ask where they got the corresponding item. The answered fashion tags appeared in a special Facebook gallery, and the best of the best were featured in the pages of the weekly magazine.9*

Let's look at this from another angle. The strength of social media rests in large part on its ability to integrate an organization with affiliates through conversation. Social media strategy puts a premium on the individual representatives of the organization to know, understand, and serve customers. This brings a new level of appreciation and a corresponding ability to creatively construct effective campaigns. The savvy campaign often uses a blend of both digital tools and traditional methods of persuasion and branding to dramatically grow engagement and activity. This is a paradigm shift that traditional advertising and integrated marketing communications professionals have, at times, found difficult to embrace. Old Media was also monopoly media, with the corresponding notion that "we'll tell you what we think is important for you, and you will act and react according to information only available from us." And then we will sit back and wait for your reaction, perhaps measuring it, but usually just simply waiting for activity.

Now the individual is in a position to know as much as the experts. Social media puts a premium on the synchronizing of an organization with its customers. In order to do so, it calls for the organization to play a number of roles, not only marketing specific products and services but teaching and nurturing the consumer or affiliate into mutually beneficial conversation. This takes work, a whole lot of work, and a continuing and complex set of technologically mediated activities that engage both the affiliates and the organization.

Underlying all of this is the role of "*secret lover.*" In emphasizing love, Homer Simpson is wise beyond his two-dimensional years. Social media engagement relies upon an underlying feeling of ease, of the belief that an organization actually understands and cares about both the present and future of its affiliates. In this, all communications by the organization must have at least a faint scent of love, of respect and caring.

Savvy social media as *secret lover*: Not easy to do when you consider that social media, when unpacked (TWA--Trendy Word Alert!), is thousands of communications by individuals representing an organization, each separate communication having to demonstrate both respect and understanding, and a unity with the goals and branding of the company and its products. Messages and warmth must be uniform across the organization. No "loved by one" service representative, "hated by another."

Think about it. Social media strategy demands internal discipline, training and policy. And an organization is filled with secret lovers *of customers.* Have you every spoken sharply and instantly regretted it? What makes a remark sharp is a momentary absence of love for another in tone and/or content. Not to get philosophical, but love, steadily and quietly, is what makes relationships succeed, drives hyper-relationships, and results in continuing and continuous engagement. This may sound simple and even a bit juvenile, but a quick read of the consulting literature on shaping business social media content finds frequent use of words like "heartfelt," "respectful," "caring," "reassuring"--great advice and the foundation of growth in both your personal and corporate life.

It's easy and hard. It's Einstein and Homer Simpson.

THE TAKE-AWAY

Social media strategy has so many variables, so many moving parts that it is easy to understand how difficult it is to implement. Successful social approaches rely both on technology and a discerning grasp of human nature to successfully establish relationships and achieve organizational growth. This is a technological age of innovation, when every day gives us new tools and applications; if you can think of a new approach to involve affiliates, the chances are the technology exists or will shortly to implement your strategy.

All of it relies upon connecting the individuals who represent an organization with its affiliates. Social media success depends upon savvy individuals and savvy organizations who sincerely integrate engagement, understanding, and love into their goal-oriented conversations with consumers. Again, so many moving parts, demanding more in the way of discipline and understanding from organizations than ever in the history of marketing communications.

But the rewards that come from knowing and understanding customers in real time are worth it. Strategies can be adjusted on the fly; issues can be addressed before they pull down an organization; and the strategic learning curve is shortened, with productive strategies emphasized and the unproductive eliminated, and growth, often extraordinary growth as the result.

Ever watchful, ever putting in the effort, ever engaged, ever in relationship, ever patient, ever loving; it's more like a great marriage than a business relationship. Savvy social media marketing is never easy, but it *is* dramatically rewarding.

Our conclusion again: Woo-hoo!

(ENDNOTES)

1   http://www.theaitgroup.com/00_competitive_farsight_resource_library2.asp

2   http://dailycandy.tumblr.com/

3   Beverly Macy and Teri Thompson, The Power of Real-Time Social Media Marketing,: How to Attract and Retain Customers and Grow the Bottom Lin in the Globally Connected World, New York: McGraw Hill,   P. 26

4   http://thechronicleherald.ca/artslife/96475-musician-carroll-parlays-his-united-experience-into-book

5   Ibid.

6   http://www.facebook.com/sarahpalin

7   http://www.sarahpac.com/

8   http://adage.com/article/special-report-book-of-tens-2011/ad-age-s-book-tens-social-media-campaigns/231498/

9   Ibid.

CHAPTER 10

PREGNANT...
WITH SOCIAL MEDIA

INTRODUCTION

SOCIAL MEDIA STRATEGY IS NOT JUST IMPLEMENTED. RATHER, BOTH STRATEGY AND CUSTOMERS ARE NURTURED, NOURISHED, AND BIRTHED THROUGH A PROCESS THAT IS NOT UNLIKE PREGNANCY. My wife brought two girls into this world, each of them different in nature but both receiving the same care and nurturing for the first nine months of their lives inside the womb. She patiently brought them to term through devotion and attention, careful to think, do, and live in such a way as to increase the chance for each girl to come into the world happy, healthy, and ready for the challenges and opportunities of life.

This chapter introduces a framework for understanding the creation and nurturing of a social media strategy that is similar to the holistic approach to birth used by my wife and so many other savvy women in bringing their pregnancy to a happy and healthy term. Social media capabilities are best approached deliberately and organically as marketing strategies, that are also "birthing strategies," two long-term two-way customer relationships. Social media strategies, like pregnancy, are the beginning of an intense relationship that requires constant listening, acting, reacting, and emotional commitment. Pregnancy is a lifetime mother/child relationship commitment; a social media strategy, at its most powerful and effective, is a lifetime commitment on the part of an organization to its customers, donors, or members.

A mother's commitment to nurture, coach, listen to, and to feed a relationship that will benefit both mother and child can be compared to the intensity and commitment demanded by savvy and successful social media strategists. Affiliates of the organization (that's what customers, donors, and members are these days, individuals in complex relationship with a variety of organizations and causes) are in need of 24/7 monitoring and care. Connections are digital and therefore dynamic, real-time, and shaped by continual conversations. A successful social media strategist regards affiliates in the same way a caring mother regards her baby, a blessing to be loved, shaped, and nurtured from the moment of conception through birth and beyond. We use the acronym TERM to frame this approach to digital marketing communication.

TERM consists of the following elements: Talk, Engage, Re-engage, and Monetize.

Talk: Encourage the right kind of *talk*, the kinds of conversations that promote greater affiliation and liking, always liking, for overall greater affinity between the organization and its affiliates.

Engage: Affiliates are not simply targeted. The savvy social media marketer seeks to *engage* each individual, guided by the benefits offered to both the organization and individual, by developing a healthy relationship.

Re-engage: The modern marketplace, shaped by digital media, is one of continuing conversation, and continuing conversations demand that we *re-engage* customers through loyalty programs and constant communications that build affinity.

Monetize: We don't just do, we do to build value, to *monetize* a relationship where those who are engaged experience real and valued benefits. After all, everyone responds to incentives that promise greater rewards and benefits.

ALL YOU NEED IS LOVE

Love is as central to savvy social media strategy as it is to life. As the great philosophers, The Beatles, put it:[1]

There's nothing you can do that can't be done.

Nothing you can sing that can't be sung.

Nothing you can say but you can learn how to play the game

It's easy....

All you need is love, All you need is love,

All you need is love, love, love is all you need.

The key to the growth and effectiveness of the Social Media category is its focus on the loving relationships of what we call *affiliates*, what public relations professionals call stakeholders, consumer merchandisers call customers, and non-profits tag as donors. The driving force behind the effectiveness of social media is the almost seamless integration of these tools, on behalf of an organization, into the daily relationships, activities, and emotions of the average individual *affiliated* with it. "All you need is love." There is something in all of us, in all of humanity that wants to connect, needs to connect, and loves to connect. We were designed with the urge for love and connection. Social Media, at its best, is the "love" category of marketing. It develops a connection centered on love: love for others, love for organizations, love for products, love for charitable acts, love for churches, love for, well, feeling love!

Such is the power of constant digital connection, linking to the right individuals, in the right way, at the right times. Just as that great philosopher, Homer Simpson said, "Donuts, is there anything they can't do", we can summarize the potential and possibilities for social media in our lives and careers: "Social media, is there anything they can't do?"

PROCESS: BORING BUT NECESSARY

Most likely you've heard it all before, but it's worth repeating: marketing and promotions are best approached systematically as a process. The Social Media category of promotional tools is similar to other marketing communications principles and practices in that it is effectively used with a five-step process:

1) Define your objective

2) Understand the operating environment of the organization

3) Craft a strategy that ties together your objective, operations, and approach in a particular environment during a specific length of time

4) Use this strategy to select and then implement the tools and tactics needed to achieve the objective

5) Create inspection routines, in which operational and marketing benchmarks are inspected and assessed, followed by adjustment (and re-adjustments).

The goal is to: systematically pursue present and future results that will increasingly bind a company and its affiliates while delivering growth through the increasing intensity of beneficial relationship for both.

However, Social Media is different in that it uses, as its foundation, the human need for a positive emotional connection with his or her world as represented by individuals, organizations, and institutions. The result is a more intense relationship than that developed by traditional media (also known as Old Media, while social channels are considered New Media).

FAQ 4: Old Media, New Media—what's the difference?

Old Media is a descriptive term that includes the mindset inherent in traditional media organizations that view communication with audiences as one-way, with few opportunities to talk back, connect, or express preferences. Indeed, in the Old Media view, consumer silence is golden. Old Media includes traditional channels such as newspapers, television, radio, and magazines, to name a few. They were and are consumed, while social media are absorbed and integrated into the lives of both organizations and individuals. Old Media are notorious for ignoring the wants and needs of consumers, creating products that invariably decline in market influence and significance. The newspaper industry, for example, is on its deathbed, having scorned readers and advertisers for much of the last half-century.

Old Media are passive, in that involvement skims the surface of our inner lives, its content largely fodder for conversation. A viewer sees something interesting on the news and tells another about it: "By the way, did you see what those nut-jobs in San Francisco did today? A bunch of them marched on government offices, insisting that leather be banned inside city limits because making leather involves killing animals. They want the city to set up police checkpoints to keep all vehicles with leather upholstery from entering San Francisco!" In the same way, strategies for traditional media emphasize getting someone to notice something or think something period, the end.

Traditional media promotions tend to be episodic, with defined beginnings and endings. A retailer wants more people to know about its factory outlet store, so he buys billboards along a 50-mile stretch of Interstate 95 for a year. A mall experiences slowing sales and stages a series of public events leading into the Christmas shopping season. The owner of an independent coffee shop is visited by a representative of a local radio station, which is offering a three-month special for 30-second spots, sold and on the air for three months.

But social media? It's a relationship, baby!!

IS SOCIAL MEDIA REALLY BETTER THAN OLD MEDIA?

In a word: Yes. The Social Media category encompasses a set of tools that can be extraordinarily effective, more so than any of the marketing communication tools previously available. Social Media is not just another category of media (i.e. radio, cable, magazines, etc). Rather, it is a mediated extension of an individual, dramatically expanding his or her ability to purposively connect with others and foster more intimate relationships. Individuals who connect through social media experience relatively deeper ties to others, engaging in more effective relationships.

In other words, social media are powerful stuff, and the Social Media category is…well, powerful stuff. Through it an organization extends its connection to individuals, experiencing greater intimacy and building community. Because social media are quickly becoming ubiquitous, Social Media is becoming the major element in the strategic plans of organizations as they seek more effective ways of creating and managing relationships with customers.

An organization does not exist without customers and the Social Media category is rapidly becoming the primary means by which twenty-first century organizations create, increase, and keep customers. It is the single most effective set of tools that organizations use to grow in the digital age. It combines the scope of mass-mediated marketing with the power of one-to-one relationships. Social Media offers the possibility of fostering the kind of loyalty for an organization that was once only possible between individuals at the family or friend level.

TALK: FROM "BY THE WAY" TO BTW

Traditional or Old Media content and consumption are the stuff of transactional and often shallow communication. Transaction: "I'll tell you about my favorite show if you tell me about your favorite show. Yes, that *Star Trek* retrospective was cool, by the way, how did you like *Jerseylicious* last night? Old Media stimulates brief exchanges of information, but rarely involves or engages. More often than not it is simply, "I'll give you my piece of information and in return you will offer your information nugget. I'll tell you who placed second on last night's semi-final of *American Idol* and in return you tell me about the hot guy chosen by *The Bachelorette* for her second date." It involves time but does not often involve the individual emotionally.

Old Media and old marketing offer a series of "by the way" statements. "By the way" shapes future conversations. "By the way" is a form of "Did you hear…" You watch a television show and tell someone about it. You read a magazine and pull from it an information nugget for use during a conversation. You rent a DVD and tell your friend about it. Shared information, but largely lacking in depth, involvement, or emotional attachment.

In the world of Old Media, "by the way" is a transition from one information nugget to the next, allowing one person to communicate a piece of information to another. It is transactional in the sense that one person sees or reads something and passes it along to another, who then provides an information nugget in return. For example, you tell a friend about the news item originating in San Francisco, where some residents are halting cars with leather seats at the city limits, cow skin being an affront to Bay area animal lovers. Then your friend responds with another piece of information of equal or greater value in transactional fashion, "Hey, that's nothing! You ought to see what they did in Las Vegas; they're demanding that the casinos prohibit customers wearing leather pants from gambling. They're making them remove their pants at the door!"

Social Media is a whole different ball of wax. It not only provides the stuff of talk but serves to create and build relationships. "By the way" (BTW) is used differently in a social medium like the smartphone. For example, you are texting back and forth with a person, and then use the intimacy of the conversation to allow him/her a peek into your inner life: "BTW…I'm here for u," "BTW…I really don't like that place," "BTW…preacher is a snoozer." BTW is often, but not always, used to intensify a relationship, to further the intimacy of individuals in relation to each other.

Relationships between individuals are about emotions and feelings, loyalty and love. Now technology and media have evolved and spread to the point that the tools of the Social Media category exist for an organization to become as much a part of the life and experiences of an individual as family and friends. This provides the opportunity for organizations to participate in both the inner and outer lives of their affiliates (customers and employees, shareholders, vendors, etc.) all while drawing closer to the individuals in the communities that are in constant orbit around the organizations and individuals, the ones creating that Virtual Relationship Cloud!

I LIKE YOU

Social Media is a different type of marketing communication category. To be worked effectively, it demands that you approach it on an emotional as well as functional level. The often dramatic results of the category depend upon building affinity, creating a bond between the organization and the individuals or affiliates with which it deals. This bond is its strength, expressed as affinity or "liking" for an organization and its services, its employees, and its products and beyond, into engagement. It all begins with "liking," which is not just an activity popularized on Facebook. Rather, it has long been the primary indicator of effective marketing programs by consumer merchandisers. In consumer marketing research, researchers have consistently found, over the past half-century, that the primary determinant of a positive response by consumers to advertising, packaging, promotions, and publicity—all of the many marketing and communications techniques—is "liking." An individual who "likes" a brand is positively and emotionally attached, and buys its associated products. For example, an individual "likes" a charity and donates, "likes" a church and joins, "likes" another person and becomes friends, and so on.

The emotion of "liking" is a *linking* feeling. In the Old Media and old marketing days, we called it branding. In the digital arena, the term *branding* is like wearing a pair of size 7 shoes when you've hit size 9 and are still growing, it's restrictive. "Liking" is so much more than branding. The "liking" of a woman for a pair of Steve Madden shoes is extended or linked to other shoes produced by the organization, thereby extending the relationship and developing a consistency in the *Virtual Relationship Cloud* surrounding the company, its products, stores, websites, and operations, a critical element of branding in the era of social media. In the same way, an individual who "likes" another person also tends to "like" the choices that person makes, the brands he or she buys, the church he or she attends, even extending to that person's family and friends. The result is a lengthening string of "likes," ultimately leading to a more intense liking that extends into loyalty and intensely positive expectations and understanding, love and advocacy.

For example, you like Jennifer, she's a good friend that you respect and she approaches life with a level head and enthusiasm. She recommends that you try the new yogurt dessert place in town. "It's better for you than ice cream," she explains, "they are so friendly and have such a really good selection! Try it, it's great!" You not only try it, you enter the store predisposed to like it, to find something to be favorable about *because* Jennifer likes it and you like Jennifer. Awareness Social Marketing Software refers to this as the "Liking Principle," in that individuals "are more likely to buy items that friends have recommended or that movie stars/sports figures have endorsed.[2]

Liking, when it is nurtured and cultivated, leads to love – for a product, retailer, charity, or church – and a relationship that deepens over time. This deepening is central to social media strategies, which are effective, and more effectively *monetized* when individuals are *engaged* and then *re-engaged* using *talk*. Hence, growing positive relationships between an individual and an organization is critical to marketing success in the new social age of marketing communication. Achieve this and you also achieve more buying, more joining, and more donating at less cost as the growing loyalty that results from effective social media strategies increasingly predispose affiliates to more effective activity in relation to your organization. How?

Let's go back to the notion of approaching social media campaigns in the way caring women approach pregnancy. You use both *nature* (the natural characteristics of your target audience) and *nurture* (your ability to create and promote healthy two-way conversation in relation to your organization and its activities) to develop relationships that produce healthy and increasingly valuable actions and reactions. Social media campaigns, unlike marketing strategies that rely on traditional media and communication channels, involve simultaneously long-term and short-term strategies, analogous to raising children. It is a continuous effort as you work to get them through the first nine months and then work to get them through life.

In Old Media marketing, you create and implement a campaign, assess the results, and you're done. With Social Media marketing, you create and implement a short-term and long-term strategy, assess the results, and apply this to continuing and new strategies, assess the results and apply this to continuing and new strategies. You are continually generating and acting upon feedback. Like raising children, a social media strategy is a continuing effort and commitment.

Let us extend the metaphor of pregnancy and apply the notion of both *nature* and *nurture* to social strategy development. The savvy social media strategist understands that there are certain natural tendencies shared by all of humanity as part of a broader design for the universe. To the extent that he or she strategically accounts for those tendencies during campaign planning and implementation, a social media approach will be exponentially more successful than traditional marketing communication methods. These designed tendencies, reflected in human *nature*, can be developed and built upon by strategies and tactics that appropriately *nurture* conversations, relationships, and responses that advance the goal of the organization in linking to its affiliates.

The natural tendencies that are particularly relevant to the birth of the organization/affiliate relationship and its subsequent nurturing include:

1. LEADERS ARE FOLLOWED. People follow individuals viewed as expert or knowledgeable. These individuals can be friends, family, or colleagues and they share an authority that lends credibility to their opinions. We as individuals do not know everything; this bit of human wiring helps us to act in areas about which we are not expert. It also allows us to make decisions about matters that we have neither the time nor inclination to dig deeply. I take particular pleasure, for example, in eating a Cobb salad at lunch; however, I don't have the time to find the best restaurant for salad in each of the many cities I visit, and rely on the recommendations of friends and/or those perceived as Cobb-salad-knowledgeable to point me in the right direction. Social media allows me to digitally bring the expertise of others into the lunch location decision. Carol King, another great philosopher, explicates this principle it in her classic song, Where You Lead: "Where you lead, I will follow, anywhere that you tell me to…"

2. OPINIONS MATTER. Popular activities interest people in part, because they are popular. People like to be part of something larger, connected to a bigger something. This works in every area of human life. Let's take sports, for example: I'm a New York Jets football fan and enjoy being with others unfortunate enough to 'like' the National Football League's most consistently mediocre team. We value the opinions of others, pay attention to ideas and activities that attract followers, and tend to attach value to that which attracts a wider audience. Why? Because we—all of us—are built this way.

3. GRATITUDE—I CAN'T HELP IT. You can't get something for nothing? Nonsense! You *can* get something for nothing: Gratitude along with all of the activity and connection that follows from it. That's why one of the most powerful words in the English language is "free." People *respond* to offerings of worth, freely given, with nothing asked in return. When something is offered at no charge and with no strings attached, a vital part of our nature generally (but not *always*...more on that later) kicks in: Gratitude. It's not "You scratch my back and I'll scratch your back" (although in some situations that is, indeed, the case). Rather, it is "Wow, you scratched my back and now I *want* to scratch your back."

Again, it is how we are designed, reacting to a gift, freely given with a greater willingness to continue in a relationship with the gift-giver. It's an intelligently designed response, the urge to connect, to be open to the undemanding kindness of another. We find this repeated again and again, from the relatively trivial and pleasurable, "Try this free smoothie." To the vital and life-changing, "Hey, look at this free Bible that guy gave me—maybe I ought to take a look at it." We are thankful for the gift, for the opportunity, for the product or service offered with no strings attached. And then we respond to the gift through gratitude:

- "Well, I sure do appreciate that free perfume sample. Maybe I'll be a bit more open to trying another product from the company's line."
- "Wow! The volunteers from a mission group just gave me this cookie. You know, maybe I'll add them to my donation list!"
- "Look what came in the mail, honey! A free DVD of the new Christian movie 'October Baby' from that church on the other side of town. You wanna give them a try this Sunday?"

However, while there is no guarantee, generally kindness or graciousness increases the chances of a favorable response. This translates into social media strategies that stress giveaways, click-throughs, and a myriad of other tactics including a deliberate strategy of conversational praise, by the organization, to increase the chances of growing affinity through favorable responses.

The *nurturing* tendencies that make social media strategies so powerful include:

1. LOYALTY GROWS THROUGH AFFINITY. It is possible, through social media nurturing strategies, to grow the affinity of large groups of affiliates (audiences, customers, etc.) to loyalty. In other words, we may develop a positive emotional connection into a predisposition to join an organization, buy products, and/or donate to a particular non-profit. We are wired to respond to nurturing strategies and the Social Media category, unlike traditional media and other forms of marketing communication, has at its core the use of nurturing to stimulate customers. The more intently and longer we like something, the more loyal we become to it. This is part of social media nurturing, instilling an expectation to be treated as special.

2. TRUST BREEDS ACCEPTANCE. For once, one of those academic words—iterative—fits, as it is possible to nurture acceptance through an *iterative* conversational process. To describe something as iterative is to emphasize its repetitive nature. We are wired to establish and appreciate routine in our lives and repetition is integral to establishing routine. Regular conversations and interaction with an organization by an individual promotes trust on both sides, establishing a repetitive and regular relationship. Reactions by each to the other become predictable and normal, predisposing each to accept the positions and assertions of the other. It works the same way that a cup of coffee with a friend at the local Starbucks, repeated often enough, brings with it feelings of trust and a greater acceptance of the views and actions of the other. For an organization, putting in place a social media capability and nurturing conversations, to the point where interaction is regular and expected, breeds trust and acceptance in the market. The Virtual Relationship Cloud becomes warm and welcoming.

3. LOYALTY AND TRUST INCREASE INTENSITY. It's all about attachment. We get attached to people, things, and animals; it's just part of what it is to be human. The savvy social marketer understands this attachment, when you build on effect and – in the words of celebrity chef Emeril Lagasse – "kick it up a notch," you increase the intensity of the relationship. This is when "more" happens – more buying, more donating, more liking, and more participation – the activities surrounding an organization/affiliate relationship that has been nurtured into intensity.

THE TAKE-AWAY

Social Media derives much of its power from its reliance upon the natural desires of individuals to connect, to not be alone in the universe. Humanity has a natural urge to connect with and care about others, at the same time responding to activities that build trust and relationships. Social media rely upon natural instincts and, consequently, are that much more effective and powerful than traditional media and marketing. Social media belong to a new and exponentially more powerful marketing communications paradigm, that of caring and sharing, nurturing and shaping, relationships that build understanding and…love. Unlike the traditional, episodic Old Media and Old Marketing approaches, social media— used correctly—allows an organization to create affiliate relationships through a nurturing relationship. Savvy social media strategies are similar to the relationship between a woman and her children, from pregnancy to adulthood.

For the first time, marketers have the ability to be continually involved on a mass scale in the lives of the individuals who comprise an organization's community. But to do this, social media must be used deliberately, choosing carefully the strategies and tools used to involve customers and prospects. Organizations must put in place goals that guide the use of social media strategies, all with the direct intention of growing affinity into loyalty on the part of the customer toward the organization and the organization toward the individuals. The individual shows his or her loyalty by becoming increasingly involved with the products, services, and/ or ministries of the organization. The organization shows its loyalty in the types of services and products it provides, shaping both current and future offerings and activities to conform to the wants and needs of its nurtured and growing community.

(ENDNOTES)

1   http://en.wikipedia.org/wiki/Beatles

2   "Social Commerce Lessons: The 6 Social Principles That Increase Sales," white paper, Awareness Social Marketing Software, February 2012.

CHAPTER 11

STRATEGY:
SOME CUSTOMERS ARE BETTER THAN OTHERS

STRATEGY

INTRODUCTION

AW, C'MON MAN!

Those of you who watch and listen to sports are familiar with this phrase; the commentator rips it off when a professional athlete says something that's especially dumb, arrogant, and/or self-serving. When I introduce the notion to people that we have to focus on some individuals over others, that's often the first reaction I get. Keep in mind, I warn, that not all customers are actually equal in value to your organization. Not all are meant to be, or should be, affiliates of an organization, that includes non-profits and religious organizations.

Yes, religious leaders should welcome everyone with appropriate beliefs to their membership-driven organizations. However, keep in mind that resources are finite—especially time, which when gone, is gone—and when you spend time on one person, you're not spending time on another. At some point you have to make a decision as to which individuals are of most value to your organization, and put your time and energy there. Then, in this age of political correctness, comes the pushback: We welcome everyone to our religious family because everyone needs what we have to offer. "Aw, c'mon man," they say, "everyone needs God."

I smile, take a deep breath, and acknowledge that everyone needs God and his love. Then I hit them with the "yeah…but." Everyone desires to feel loved and included, but you can't embrace and love everyone with *equal success*. Some people may be simply too hardheaded to accept what you're offering; others may not like your approach; and still others have different priorities. Some don't believe in God, others doubt his existence, and still others are of a different faith. You can't market to everyone with equal intensity on your part and achieve equality of results, just like you can't embrace everyone, love everyone, be interested in everyone. You get the picture. You have only so much time on this earth and it is impossible to satisfy, love, and/or embrace *everyone*.

So you have to choose the type or types of individuals that offer the best chance of a successful relationship with your organization. In other words, you choose the segment or segments of the market that you will target in your social communication campaigns. Describe the target in personalized and specific terms; bring the individuals, who represent your market, to life. Your operational and marketing objectives are then customized to maximize the opportunities created by engaging with the many individuals represented by that person. This group becomes your *target* or *targets*. The simple act of targeting an individual with certain defined characteristics shapes your marketing and operational approach, puts your customers in a utilitarian light, and lays the groundwork for successful social media marketing.

Once you know where you're going (organizational goals) and the type of person you want your organization to have a relationship with (target), you have increased your chances of creating organization/ affiliate relationships that are conversational and long lasting. Sounds like common sense, something that you know intuitively, right? You would be surprised at how uncommon common sense actually is, just look around you. Organizations forget that they are individuals, and individuals tend to bond more with certain types of people as opposed to others. Despite sounding almost philosophical, it actually makes sense. All digital initiatives begin and end with the individual most likely to relate to other individuals in hyper-relationship with your organization. Your target is that slice of the market that offers the greatest opportunity for mutually beneficial relationships.

This chapter outlines the marketing worldview behind social media marketing; it is a selective and pragmatic view, borrowing the best of traditional direct marketing and marrying it with the new and ubiquitous world of socially networked media. Particular customers who are engaged through social media, more often than not, prove to be an organization's most productive segment of the market, also known as "BFFs" or Best Friends Forever. These customers are brought into confluence through social media strategies with the company, its products, and functions by participating in both continuing and episodic conversations. As a result, the targeted segments are more likely to buy, give, donate, and/or join.

Confluence is one of those words we don't come across every day (unless you're a geologist examining the point where two or more streams are coming together and you stroke your chin and exclaim, "Ah, there be confluence!") but we've borrowed it from the geologists. In the Social Media World, we use it to refer to the technological point where different flows of customers are brought together with the company to share meaning, interest, and activity.

Organization and affiliate relationships are fueled by the emotional pull of conversations that benefit both parties. This is not the stuff of Kumbaya, a generalized feeling of goodwill that many people claim will bring world peace and free Frappuccino's from Starbucks. Rather, it is a complex digital merging, of the affiliate and the organization into a relationship, rooted in the organization's understanding that some individuals fit its continuing conversation better than others. When compared to traditional marketing communications, a key reason for the comparatively greater success of social media strategies is its personalization, of an organization for a customer or constituency, and the foundational assumption by marketers that those who involve themselves in relevant and interactive conversation become more active affiliates.

In other words, some individuals are more valuable than others for an organization, which means they should be addressed differentially, with more emphasis on those who will provide the greatest return on your social media investment.

BACK TO THE FUTURE

In *Back to the Future*, a science fiction classic from the eighties, teenager Marty McFly is accidentally transported from 1985 to 1955, where he meets his future parents and accidently causes his future mother to dislike his future father. When he returns, he finds his life has dramatically changed for the worse. He travels back again in the time machine to repair the damage done on his previous trip and get his parents to fall in love, thereby returning the future to normal. So, what's the point? First, I enjoy movies and working them into a textbook is rather natural for me and loads of fun.

Second and more to the point, successful use of digital interactive media is similar in approach to all of the great marketing strategies of the past. If you want to know how to increase the chances of social media success in your organization, you have to go *Back to the Future*, two or three or even four times, whatever it takes. There is much the past can teach us if we're willing to listen and apply its lessons.

First, the great marketing successes of the past, present, and future all have one thing in common: a defined target. A target may be a single segment, or include several segments of a larger market. Great marketing begins with a clearly drawn picture of the type of person or persons that are most likely to join, donate, or buy. The Whole Foods grocery chain has climbed to marketing stardom on the backs of shoppers who describe themselves as, "…modern eco-foodies (who) are full of good intentions. We want to save the planet. Help local farmers. Fight climate change—and childhood obesity, too." They love everything from "recyclable cloth bags to Michelle Obama's organic White House garden."[1] Meanwhile, another grocery superstar, Costco Wholesale warehouse has achieved industry-leading growth by catering to bargain hunters who appreciate its "sparse layout" and "no-frills shopping experience."[2] Dramatically different approaches by two businesses have yielded success in the same industry. Their social media use follows the same patterns. Whole Foods shoppers enjoy keeping in touch with the firm through its website, devouring the lifestyle content, and taking pleasure in discussing the latest gourmet tofu recipe or tweeting the arrival of a new brand of granola. Meanwhile, Costco shoppers scan their emails for the latest bargains, and flock to the store for great deals on a wide variety of common household goods. Although the presentation of each grocery superstar is different, each succeeds because they are targeting different market segments of individuals with unique lifestyles and consumer behaviors. Key word: unique. The savvy social marketer begins with the unique character and behavior of the preferred target segment and then works back to and through the interests of the organization.

Second, all of our social media marketing and operations rest on a foundation of consistently and passionately touching the individuals in our selected market segment. This happens through an engagement process that requires the organization-to-consumer-and-back relationship to be defined by *TERM: Talk, Engage, Re-engage,* and *Monetize.* All digital interactive strategies require taking the market from where it is, as gauged by the conversation during a 24/7 social cycle, to where we want it to be through constant, directed conversation. Remember, the social, digitally-powered market *Talks* throughout the week, not restricted to when the charity, church, or department store is open for business or a promotion lands in the customer's mailbox, and the topic of the moment can include any happening or feeling from the past week or month involving any individual or organization. The individuals that comprise the market expect to *Engage* with organizations digitally, through a variety of social media applications, on numerous devices like smartphones, and they expect to be *Re-engaged* through conversation by the very same organizations. Everyone expects something of value from engagement such as discounts, an invitation to a frequent buyer club, and ultimately, feelings of love, all the stuff of conversation. Finally, we *Monetize* a slice of the market (our target) through the addition of tangible value to a set of activities, something that is of measurable benefit in some way to an affiliate in relationship to our organization. All strategies should offer something of value to a targeted segment.

WE SHAPE OUR TOOLS AND THEY SHAPE US

"We shape our tools," Canadian English professor and media guru Marshall McLuhan said, "and then our tools shape us."[3] His ruminations on media effects became quite the fad among educators and amateur philosophers, during the last part of the twentieth century, as they hailed his head-scratching non-sequiturs as wisdom for the ages. "We don't know who discovered water, but we know it wasn't the fish," he would proclaim, and academic audiences would erupt in cheers. Water, fish, wisdom for the ages!

Yet, if we look beyond the fish and the preference of highly educated audiences for the nonsensical, there is a truth here; you cannot separate the digital media ecosystem from its human users and vice versa. Who we are shapes our choice and use of social media, which exist to facilitate human communication in relation to an organization or community. Social media are, for the most part, everywhere and easily accessible so that we may, with a few barriers, express even our slightest desires and whims. However, while the choices grow exponentially every month, organizations and individuals are limited by the fit of their characteristics and goals. A corporation, for example, is limited in strategy by the preferences and characteristics of its target market segments. Individuals representing the company may say whatever they wish, but its targeted segment may or may not listen.

A company that is in business-to-business (B2B) market may choose (because of the no-nonsense nature of its customers) to use Facebook as a hub, with its great amount of corporate customizing tools, and LinkedIn as one of its spoke media tools. Meanwhile, a church may use its own website as a hub while using ChurchCloud, Twitter, and rich media email blasts as spokes. The social media strategy developed by an organization depends upon the nature of the targeted segments, and different segments have different preferences; it's that simple!

It is important to remember that social media are simply aggregations of individuals in digitally mediated conversation (an academic insight, folks). Modern digital media are enabling, in that, they allow individuals to share information and views as they come to the forefront on a particular topic or activity during the course of life. The savvy social media marketer chooses digital media with a specific type of individual in mind. A recent survey of major brand owners found that more than three-quarters are changing their marketing to better serve the unique communications flowing through social media from consumers.[4] Social media allows these brands to better differentiate the characteristics of their best customers (the brand targets) from the mass of individuals of lesser importance to them in the marketplace. As one social marketing expert explained, the continuing exposure to targeted customers produces learning "from our interaction with people via reviews and online forums (that) really help us understand what consumers…feel are the really pertinent issues around" their products.[5]

Savvy corporations spur growth through understanding how the communications of their best customers reflect meaning. General Electric, for example, has modified its approach to research and development because of the growing understanding of the differences between its loyal customer segments and those more likely to switch to another brand. "As gadgets are getting more mobile, and more powerful, we have more information to process than ever before," the company's chief marketing officer explained.[6] He is using the social media conversation flow to understand the "stories" consumers tell about GE products and staff and redirecting a significant portion of their research effort for this purpose.

Different strokes for different folks, it is surprising how many people forget this critical insight underlying all social media. It takes a certain type of person to bond with another person, just as it takes a certain type of individual to bond with a certain type of organization. A social media strategy is organic, in the sense that, it flows out of the strength of the uniqueness of the chosen targeted segments. Certain individuals are more attracted to certain organizations than other individuals and, consequently, are more *attractive* to those organizations.

We choose our social media tools to fit our targets and their styles of conversation. In this respect, it is important to consider the "stories" (as GE and so many corporations put it) that your affiliates, who benefit most from association with you, are telling. Again, the focus is on your targeted affiliates, not the many who are there or wander into range but for whom there is little long-term opportunity. This is in line with the bit of business wisdom that we've discussed before; 20 percent of your customers do 80 percent of your business. This is fairly widespread human behavior. Some of us are more enthusiastic about a particular activity than others. I, for example, stop at Starbucks about once every two weeks; a co-worker has a twice-a-day $200+ a month habit. Here's another analogy, a group of people sit down to eat pizza; most eat two or three slices, while a smaller number eat less and another few eat six or seven slices. A few people, then, produce most of the eating and, later, presumably consume the greater amount of antacids.

Who we are and how we interact with the world determines (1) whether we are a worthy target for affiliation with an organization, its products, and services, and (2) what social media are strategically used to bind us to the organization. The stories we tell, the conversations we have, are a function of who we are and how we want to be perceived, our different personalities and characteristics. Some social media tools are better than others at working with specific targets. Following directly from choosing one individual over another is the notion that some tools are better than others in communicating with them, as social media use reflects a more generalized approach to life on the part of an individual. Bottom line: there are few surprises when it comes to social media use, just as there are few shortcuts to effective social media strategy. It begins and ends with the characteristics of your target, which should reflect the thinking of your organization as to the most likely prospects for relationship. Talk to me all you want, send me offer after offer, but the likelihood of my joining thirty experts for a foreign policy discussion cruise to Alaska ranks somewhere down there with signing up for a VIP tour of Public Broadcasting System headquarters in Washington, D.C. However, offer a trip to Disneyworld or free barbecue at an award-winning rib joint and I'm all yours! Heck, I got married in part because of the persuasiveness of a German chocolate cake.

The savvy organization that uses social media considers the differences between the most likely users of its services, buyers of its products, members of its congregation and those who use, join, or buy from others. The strategic focus of the organization is specific rather than general. The savvy social user focuses like a laser on specific customer benefits, specific customers in relation to a specific campaign, and specific strategy. General studies of social media use, while interesting perhaps, are less than helpful. One such study, for example, has shown: [7]

- The more extraverted you are, the more likely you are to use social media.
- The more stable you are, the more likely you are to engage in conversation with others
- Younger adults use social media
- Those open to new experiences use social media

No duh! And you were bored to tears by that communication theory course? There are characteristics of people in general and then there are specific characteristics of individuals who are most likely to affiliate with your organization; the latter offers the most value for savvy social networking. As we discussed, what's good for *everyone* is not good for a *specific* someone in relation to a *specific* organization. Savvy users of social media continually monitor the socially networked uses and conversations of affiliates to better understand the uniqueness of an organization's best customers/donors/members and create more relevant and productive social strategies. General Electric, Unilever and Wal-Mart, for example, are among the many major "brand owners trying to enhance their understanding of digital technology and behaviours (sic), reflecting the opportunities and challenges these trends present." [8] This information is then used to define and refine strategies aimed at the differentiated target segments these companies judge the most productive for growth.

SELECTING A TARGET

This is a boring subhead, but it's boring with a purpose. You have to select a target. A target gives you something to aim at... That's called a *target*. A savvy corporate user of social media does something as traditional and boring as choosing and defining a target, the preferred set or sets of individuals. In other words, select the types of individuals who will best fit the types of relationships you envision as profitable for your organization. A target appropriately defined is a stereotypical description of the type of individual-writ-large that you want to bind to your organization and its activities. Some lifestyle targets serve as examples from the Nielsen PRIZM segmentation system:

Country Casuals:

There's a laid-back atmosphere in Country Casuals, a collection of older, upscale households that have started to empty-nest. Most households boast two earners who have well-paying management jobs or own small businesses. Today these Baby-Boom couples have the disposable income to enjoy traveling, owning timeshares, and going out to eat.[9]

Domestic Duos:

Domestic Duos represents a middle-class mix of mainly over-65 singles and married couples living in older suburban homes. With their high-school educations and fixed incomes, segment residents maintain an easy-going lifestyle. Residents like to socialize by going bowling, seeing a play, meeting at the local fraternal order, or going out to eat.[10]

Money & Brains:

The residents of Money & Brains seem to have it all: high incomes, advanced degrees, and sophisticated tastes to match their credentials. Many of these city dwellers are married couples with few children who live in fashionable homes on small, manicured lots.[11]

Some individuals are a naturally better fit for your organization than others. It is important to define the type or types of individuals that will make for a more lasting and intense relationship, understand what makes them different from others in the market, and describe them in colorful terms, making them as human and understandable as possible. The Nielsen descriptions are just three of thousands of ways to describe a target market segment. However, a stereotypical description can provide the creative foundation for social media strategy that hits the mark. And hitting the mark leads to growth.

A successful organization has a well-defined product or service and a distinctive presence in the marketplace. Most successful turnarounds of failing companies involve redefining and tightening the nature and branding of their product offerings. Dunkin' Donuts, for example, steadily lost customers until the company sharpened its focus by defining its target as everyday people who like to order without using the terms "Grande" or "Venti," even hiring John Goodman (an actor who personifies the 'everyman' image, to provide voiceovers for its commercials. This resulted in extraordinary growth for the company, which is adding more than 500 franchise outlets a year around the globe.

THE TAKE-AWAY

For the growth purposes of an organization, some individuals are better than others. The goal of all of this is growing your organization, expanding your scope and doing more of the work of the organization more effectively. And listen, always listen to the individuals who comprise your target market segment or segments because of who they are, what they think, how well they relate to your organization and its services. These differentiated groups of individuals have much to say about your organization and why they are in hyper-relationship with you simply through their activity and conversation. The savvy social media user continually gathers insight into the differences between its target and other consumer or affiliate groups, using that insight to continually refine both strategy and services.

Yes, birds of a feather *do* flock together. And our purpose is to add to the flock and thereby grow the company.

(ENDNOTES)

1   http://www.foreignpolicy.com/articles/2010/04/26/attention_whole_foods_shoppers?page=full

2   http://rockcenter.msnbc.msn.com/_news/2012/04/25/11376778-no-frills-retail-revolution-leads-to-costco-wholesale-shopping-craze?lite

3   https://www.historica-dominion.ca/content/heritage-minutes/marshall-mcluhan

4   http://www.warc.com/LatestNews/News/EmailNews.news?ID=29518&Origin=WARCNewsEmail

5   Ibid

6   http://www.warc.com/LatestNews/News/EmailNews.news?ID=29432&Origin=WARCNewsEmail

7   Who interacts on the Web?: The intersection of users' personality and social media use
Teresa Correa *, Amber Willard Hinsley, Homero Gil de Zúñiga Computers in Human Behavior 26 (2010) 247–253

8   http://www.warc.com/LatestNews/News/EmailNews.news?ID=29323&Origin=WARCNewsEmail

9   http://www.claritas.com/MyBestSegments/Content/tabs/filterMenuFrameWork.jsp?page=../Segments/snapshot.jsp&menuid=91&submenuid=911

10   http://www.claritas.com/MyBestSegments/Content/tabs/filterMenuFrameWork.jsp?page=../Segments/snapshot.jsp&menuid=91&submenuid=911

11   http://www.claritas.com/MyBestSegments/Content/tabs/filterMenuFrameWork.jsp?page=../Segments/snapshot.jsp&menuid=91&submenuid=911

IT'S UP
TO YOU!

CHAPTER 12

SAVVY SOCIAL MEDIA
HALL OF FAME:
IT'S UP TO YOU!

INTRODUCTION

BASEBALL GREAT YOGI BERRA ONCE SAID OF EMPTY SEATS IN THE BALLPARKS OF MAJOR LEAGUE BASEBALL, "IF PEOPLE DON'T WANT TO COME OUT TO THE BALLPARK, HOW ARE YOU GOING TO STOP THEM?"

The same is true of organizations: if people don't want to engage with your organization, how are you going to stop them? As even government has learned, you can't stop people from not doing what they don't want to do, not without a gun, anyway (gosh-darn the double-negatives, full speed ahead!). We end the book as we began, bringing it back to you, dear Reader. Sure, it's about the customer, but in a sense it is really all about you and how you see the world, how you add value to organizations by helping them achieve growth. And by doing so, add value to your own life and career.

Back to the Future, sequel #389: We again emphasize and end with real-world learning by example while drawing, especially in this chapter, the appropriate lessons from selected stories of success in the age of social media. We look at how one savvy organization in particular has, through the individuals who work for it and its technology engineering, related to customers in such a way as to both grow the company and also provide extraordinary benefits for its affiliates. It is a tale drawn from *best practices*.

We're just skimming the surface here with this example and the awarding of our first Savvy Social Organization of the Year Award (SSOYA). You can go anywhere online and locate thousands of articles and books that mention the success of a variety of companies and organizations that use digital tools to bring them closer to their customers. In fact, all you need do is go to the website of 1to1 Media, for example, a division of Peppers & Rogers Group, sign up (it's free at the moment of writing, an awesome bargain considering the tuition of the average university begins at a mid-priced Toyota and rockets to BMW territory in the blink of an eye and offers a product that doesn't even approach the wisdom of 1to1 Media), start reading and, an hour later, you'll have super-sized your brain with an unparalleled interactive marketing education, and ingested no trans-fats![1] (No, they didn't pay me, they're just that good!)

Socially mediated life imitates art, or so a great philosopher has said (me, actually, I think of myself now, in my second career, as a Homer Simpson of higher education). So we turn to great performing art to draw lessons using the approach of another of the savviest individuals I know, Leroy Jethro Gibbs, star of one of television's longest running shows, *NCIS*. Do I really know him? Of course I know him, I've spent the last decade or so watching him put a variety of very bad individuals behind bars when he isn't killing them in this crime drama. He occasionally offers rules that also serve as life lessons, preparing you to pick yourself up and, as Larry the Cable Guy (whose Ph.D. in Life is the equal of anything the Ivy League can throw our way) admonishes, "Git'r done!"

SAVVY RULES! *NCIS* MEETS SPOTIFY

Supervisory Special Agent Leroy Jethro Gibbs has rules for the members of the Naval Criminal Investigations Services team that he leads on *NCIS*. These rules have guided the team to unparalleled success in solving crimes (along with assistance from the writers of the show). The rules are immensely practical (Rule #39, "There is no such thing as coincidence") and cut through the fantasies that stand in the way of successfully dealing with reality (Rule #16, "If someone thinks they have the upper-hand, break it!" For Gibbs, there's none of that "all-we-are-saying-is-give-peace-a-chance" nonsense).

We take the same approach, striving to be practical and grounded in the reality of human behavior rather than the unicorn flatulence fantasies of wishful thinking (this description is adapted from the notion of our trendy elites that your home can be powered, not by coal but by unicorn flatulence). So as we look at ways in which organizations have successfully created and continue to grow through social strategies, we tie this back to the ways in which you, dear Reader, can apply these lessons to come up with a win/win for you and the organizations you're part of. And we do what higher (and, judging from the state of public education in the United States, lower) education and so many managers (especially at the beginning of your career) rarely do: provide useful information and advice to assist you in navigating the *real* world of life outside the classroom.

Let's call this *Savvy Rules!* Creating *Savvy Rules!* For social strategy assumes that you can learn from the wins and losses of others, taking care to adapt the former and avoid the latter. Underlying all of this is the understanding that every strategy involves people. And people are the foundation of the strategic operations of Spotify, the digital music streaming service and the inaugural member of our *Savvy Social Hall of Fame*.

The upstart music service has grabbed increasingly greater chunks of the market away from more established competitors, including the 800-pound gorillas of the category, Pandora and iTunes (or two 400-pound gorillas, I suppose). It is now the fastest growing entertainment-related Internet site, gaining more than 8,000 subscribers per day.[2] Its host of services, from free streaming to subscription streaming, and range of listening and sharing options have made it an imposing presence in the digital entertainment industry. The service began in Europe in 2004, entered the United States in 2011 is growing at a greater rate than, for example, Netflix and Sirius XM, both of which have had their share of buzz.

The company has built social media links into its product, integrating with Facebook. Music preferences and suggestions may be automatically shared with friends. Spotify does not just start conversations willy-nilly (or is that Milli Vanilli?). Each *spoke* in the organization's social network is deliberately selected to reach different portions of the targeted market segments, and each is also used to provide the company with a greater understanding of how its affiliates use its products. The organization has a customer-centric culture at its core, purposively growing the hyper-relationships between the organization and affiliates, and affiliates with each other.

What makes Spotify so effective in its market? The deliberate and savvy use of advanced social features, allowing users in community to create play lists and share music, in an intuitive approach that is so very human and therefore unabashedly social. If Gibbs were alive and standing right here (as opposed to standing in my 48" LED flat screen television, where he's still alive but too small to have a Starbucks with), he would look at Spotify and say, "I know exactly the lesson to draw from this." And he would propose that the foundation, of the success, of Spotify in a free marke,t peopled by people who have choices, is this:

SAVVY SOCIAL RULE #1:
Always remember you're dealing with real people.

From Spotify executives, to its programmers, to its sales and advertising operations, all know they are dealing with real people. Spotify technology is based upon software development that assumes people use the service for the same ol'-same ol' reasons that control much of the way we live our lives, reacting to a deeply personal urge for significance and yearning for connection, for ease and simplicity in life and *love*. Spotify makes it easy to be human, allowing subscribers to laugh, cry, and be part of the world around them by integrating music into their search for joy and meaning.

EASY PEASY SPOTIFY

Spotify is, above all, *easy*. It is a product, a service, and a set of technologies that start with the customer and then works back. In fact, we borrow a turn of phrase from classic American culture to describe its service: "Easy Peasy." That little girl in the 1970's television commercial for Lemon Squeezy dishwashing liquid may have had visions of Spotify dancing in her head (another artsy bit of phrasing) when she turned to her mom after cleaning a sink full of dirty dishes and exclaimed, "Easy Peasy Lemon Squeezy!!"

Easy Peasy Spotify. Spotify users customize the music service environment in a way that takes into account individual tastes and needs. Daniel Ek, the founder and CEO of Spotify is not just a visionary; rather, he is human, profoundly human and has an intuitive understanding of the social world that Spotify affiliates (you and me, dear Reader), both present and future, are living in. It is a world where music more or less appears, where creating and publishing music is no longer a backbreaking slog through the experiential and functional swamps of the music industry, and listening to music not a complex set of transactions. Rather, as Ek notes, "anyone can record a record, even [in] their own homes"[3] and so it is Easy Peasy for independent artists to operate in the Spotify space, just as it is easy for individuals to find them. Spotify begins the path to hyper-relationship with simplicity and power: connecting individuals to their music and community, making Spotify "a social medium, where music is discovered through friends."[4] Easy Peasy. Its global operations are built in the same way. In New York City, for example, which now serves as its world headquarters, the emphasis is on building community with affiliates *rather* than bureaucracy.

Beginning with Facebook, Spotify has used social media as the foundation of its music services. The company's approach of integrating its software into Facebook provides customers with a seamless approach to listening, sharing, and discovering music. Users listen to music, which is then directly posted to their homepages and the newsfeed of their friends. The process is natural and easy, and takes no effort beyond an initial setting, social connection is automatic. Every song has the potential to be a conversation, and every conversation the potential for hyper-relationship. An interested friend sees a song that another is playing, clicks on it and goes to Spotify to listen to the song rather than Amazon or iTunes (as is the case with other music players) to buy the song. Instant gratification and no delay, Easy Peasy!

To make it even easier, there is a growing list of other applications for Facebook users who link with Spotify. Some might use a Music Dashboard, where they may discover music while seeing what their friends listen to sorted in a variety of customized play lists. Other examples include Real-Time Ticker, which shows what each individual in the community is listening to when they are listening; Just Hit Play in which an individual sees a friend listening to a particular song and may play it immediately in real time; and Musical Stories, where a conversational narrative is shaped by sharing play lists and songs. Individuals in the Spotify user community constantly connect and share through the integrated social media network of Spotify during all parts of the music and entertainment process.

This leads us to our next rule:

SAVVY SOCIAL RULE #2:
Make it easy for individuals to affiliate with you.

So many people make things hard. The cashier who ignores you at the checkout line makes it hard for you to form an attachment to a community. A candy company makes the online checkout process so convoluted that a significant number of customers quit before finishing an order. A candidate's political website makes it impossible to contact or, when you do, you get silence. The agent on the live chat of a customer service area of a website is not helpful, and somewhat sneering in tone. A university department refuses to put its degree programs online because it doesn't believe that digital education is "robust," despite the almost unanimous verdict of non-traditional students that online courses are welcome and helpful.

I could go on and on, but an organization that makes relationships easy provides its affiliates with a natural path to affinity. However, it is usually quite a bit more work to make it easier for an individual to "do business" with your organization. Retailers, for example, must put together detailed training programs for store associates who may not naturally put in the effort to provide welcoming service; E-commerce sites have to take the time to develop or acquire software that is user-friendly, adding more testing and difficulty to the programming process; a charity may have to undertake the time-consuming task of explaining their passion for a less dedicated pool of potential donors. Spotify has invested hundreds of thousands of hours in programming and processes to make life easier and more connected for subscribers. Work and more work, processes and monitoring and leaders in tune with customers, *relationships*, both corporate and personal, take work. In other words, as Yogi Berra said, "Always go to other people's funerals, otherwise they won't come to yours."

This applies both in your personal life and a company's relationship with its customers. You're 23 years old and move to a new city for a job. Do you snarl at your landlord, shake your head when someone talks to you at the fitness center of your apartment complex, and run over the toes of the valet who holds the door of your car after dining at a fine restaurant (assuming you've swiped mom's credit card)? Or do you smile, introduce yourself, and drop a few thank-you's here and there?

Which makes life easier? Be difficult and, generally, the world returns in kind. Making life easier for others usually pays off. . Another savvy observer of human behavior, Nat King Cole, offered this advice in the form of a beautifully crooned song:

> Smile though your heart is aching
>
> Smile even though it's breaking
>
> Where there are clouds in the sky, you'll get by
>
> If you smile through your fear and sorrow

Smile and maybe tomorrow

You'll see the sun come shining through for you

Savvy social media users understand this and build value by making it easier—more inviting, involving, of more benefit—for customers to affiliate with their organizations. However, it takes work to do so. Spotify understands the hard work that it takes for an organization to be easy, *savvy easy*, as it were. They have built an organization and product from scratch, making the effort to rethink many of the stereotypes regarding distribution and consumption of music. Great expense and great effort, but on behalf of both present and future customers who have and will find it less difficult to affiliate with them than competitors. It is much more work to make the customer experience easy and memorable, but Spotify has been rewarded by the market with involvement, buzz, and advocacy.

Spotify's use of Twitter demonstrates its shrewd grasp of social media fundamentals. When the music service first launched, it used Twitter to drive traffic to the Spotify website and create accounts. Twitter users continue to follow them on Twitter and use retweets as well as Twitter hashtags to promote Spotify membership. Hashtags are integral to the social media microblogging culture, used to bring users into the conversation revolving around particular subjects. Remember: communities within communities?

Spotify continues to ground its services in a savvy mix of interactive direct marketing techniques that rely on social networking to build a Virtual Relationship Cloud surrounding the organization and social networking tools. It has pioneered the seamless integration of social media into the basic engineering of its products. All of the Spotify social media *spokes* allow users to share songs and playlists, creating a continuous cycle of sharing not only the music, but also Spotify and its features in general. Talk, talk! Talk, talk! Talk, talk! The Music Machine may have been a one-hit wonder, but Spotify is likely to top the charts for many years.

Add to this another technology strategy ripped from the characteristics of its users (*Law & Order* redux): the notion of an app inside an app. Spotify users of all ages are social media natives and, while this does not qualify the Spotify community for a license to operate a reservation casino, it does provide the enterprise with a community of users and prospects remarkably at home in a digital networking environment. The Spotify *hub* application has been cleverly designed to host other user-selected apps that add features and power for the music-listening community. This approach is similar to that used by iTunes, allowing users to purchase and/or download apps by independent developers that add music-related features to Spotify.

FAQ 15: How is so much customization beneficial? Doesn't it isolate individuals and distract from the community of all music lovers?

This is one of the most misunderstood notions of community. There are communities and then there are communities. I live in a neighborhood of older people, mostly empty nesters. Some go to church, some don't. Others garden, still others don't. Some grill, others don't. I trade barbecue tips with a few but rarely go beyond a wave or brief hello with most of my neighbors. And yet when a developer threatened us, the community bounded by geography responded together, regardless of smaller special interest groups, the communities within the community.

We form communities within communities. We're different, you intuitively understand that and, because we share a single characteristic, it doesn't mean that we will necessarily be close. Spotify has done something so clever, so discerning, so, so, so savvy that they deserve our first **Savvy Social Organization of the Year (SSOY) award.** They recognize that hyper-relationship is a function of a multitude of intersecting points in lifestyle, world view, behavior, and all the other things that make us part of a larger human community, all as reflected in music choices. They allow the market to slice and dice itself in a variety of ways through affiliation with Spotify's digital products and services.

This is nothing short of ingenious. In the larger community, all enjoy an affiliation because of the product. It is easy to use, offers variety, makes it easier through a growing multitude of applications to add value, choosing music according to, say, mood or popularity, among other apps. And then there are the smaller communities within communities. Lovers of southern gospel connect with other lovers of southern gospel; teenage polka enthusiasts find other teens just as passionate about polka; classic rock is sliced and diced to highly individualized classic rock that my friends and I like. I like, my friends like, music done right, music done individual, and music done easy. Real music for real people living real lives, who'da thunk it? What does this mean for another organization? Social marketing can be extended and integrated into the very structure of a product or service. A grocery delivery service can design in social technology that assists individuals in sharing food choices with family and friends; a home crafting company can encourage, through technology, the sharing of projects and styles; the possibilities are endless, limited only by imagination.

SAVVY MARKETING IN A MESSY WORLD: LESSONS FROM MYRTLE BEACH

The Social Media category is about technology and it is not about technology. First and foremost, it is about people. And, lastly, it is about people as much as it is about technology. Spotify combines a love of its affiliates with a passion for music and passion for technology. Its business plans are based upon the reality of messy people in a messy world. Is this description meant to insult? No, it reflects a reality in which individuals who have the ability to choose tend to make the choices that best connect them to a loving and fulfilled network of similar individuals. And, because of the difficulties involved in even the simplest choices and the living of life, people do not always make the best choices. The savvy social media strategist understands this and makes it easier for affiliates to do the right thing. And so we come to our third rule, education a la NCIS:

SAVVY SOCIAL RULE #3:
Savvy social strategy involves a mix of technology, social networking, and individual benefits; no one works without the other two.

The world is messy, a difficult place to grow an organization. But an organization increases the probability of success when its strategies and tactics reflect its understanding of the difficulties involved in making choices (Savvy Social Rule #3 in action). For example, I'm writing this chapter from the 11th floor balcony of a condo in

Myrtle Beach, South Carolina with the following bit of wisdom welded to the rail in front of me:

> *In Order To Maintain An Attractive Facility, We Ask That You Do Not Feed Seagulls, Hang Towels Or Clothing From This Balcony. Throwing From Balcony Will Cause Immediate Eviction. Thank You.*

What's the point? A primary goal of the organization that manages the condominiums, on behalf of the community of owners, is operating an attractive facility that provides vacationers with a relaxing time at the beach. An attractive facility that delivers a premium experience will also be able to demand a higher rate for owners who may rent to others when they are not themselves staying, and provide a better return on investment for owners when do they sell. So there, welded to a rail in front of me, sits the plaque. This is what we want so we need you to not be a slob, not do stupid things… hey, wait, I don't live in a zoo! I may be ignorant, but I'm not stupid! Isn't this insulting?

Not really, when you consider property managers are dealing with owners and vacationers who may not always do the right thing and need to be reminded of what those right things are, and that includes me, who should know better. I'm responsible, right? Got values just like I got milk! Well, I'm also a husband and father of two girls, a father who once dropped a ball out of the window of a sixth floor condominium at another beach many years ago to show off for his toddler daughters. Missed the pool and barely missed splattering a sunbather all over the deck. Wrong, wrong, wrong, but I did it.

Why? Because doing the smart and right thing is not always apparent all the time and no matter who we are, we do and will continue to do dumb things. Dumb stuff happens quite naturally. Even the smartest and best of us need reminders, explicit instructions, explicit reassurance, and explicit material for conversation. So the goal is that the attractive but obvious sign on the balcony will join the condominium landing page, the association owners' email address list, the Facebook page, Twitter account, Pinterest, Tumblr, and private digital bulletin board to guide the owners from all walks of life who share a piece of the beach and need to be guided in community.

Great literature, like great television (*The Simpsons* comes to mind, as does *NCIS* on the television side, while *The Watchmen* and *The Dark Knight Returns* top the Great Literature charts as graphic novels, right up there with the prose greats of yesteryear, *Tarzan of the Apes* and *Gone With the Wind)* acknowledges and deals with the messiness of the real world and real people. It is difficult to grow a company. One of the greatest philosopher-historians ever, Louis L'Amour, author of more than 100 works of Western fiction, noted through one of his characters that "Victory is won not in miles but in inches. Win a little now, hold your ground, and later, win a little more."

Awesome wisdom, for both organizations and individuals, success in the marketplace comes from adding a bit of value today, building on it, and leveraging the win and its lessons to add more value tomorrow. So it goes, one step at a time, one detail in a strategy that leads to another; one tool put in place that integrates with the next; and one enthused affiliate eased into networking with another and another and another.

Adapt the worldview in this book. Take the framework for using social media tools and techniques and make it your own in service of an organization. Think application, of getting stuff done, of incremental progress and adding value for others. Yes, others: Your employer, the organizations you affiliate with, always looking for ways to add value to a larger enterprise, to the labors of others. And the rewards will come, dear Reader, the rewards will come.

THE TAKE-AWAY

The winner of the first Savvy Social Organization of the Year Award (SSOYA) is the streaming digital music service Spotify. This organization has combined social technology with a discerning understanding of the social and music needs of real people in the real world to drive extraordinary growth in the few short years of its existence. This organization offers lessons for your career. Remember: all of life is personal *and* social.

The savvy social user understands that the future will be filled with the one thing that has puzzled us in the past, confuses us in the present, and will no doubt challenge us in the future: People. Marketers don't make organizations succeed; people (as customers) do. Operators don't make companies work; people do. Guns don't kill people…oh, never mind. You get it. Spotify exemplifies the savvy approach: understand people, make it easy for affiliates to enter into hyper-relationship with the organization and its community, and keep in mind that loyalty comes when the social relationships of the community deliver personal benefits to individuals.

Above all, add value by making life easier—seamless, when possible—for those around you. Focus on results for the organizations you're part of: growth for the company that employs you. Social networking technologies are ubiquitous and fast-changing; however, to the extent that you can make those technologies and processes useful and seamless for the individuals affiliated with your organization, you'll win! Two of the most difficult parts of life are social relationships and technology. The savvy user—you, dear Readers—will use the Social Media category to enhance the benefits delivered by the mix of technology and social relationships, spurring growth through a natural integration that makes it part of the life of the individual affiliate of an organization.

All social networking technology, all marketing comes down to the benefits derived by real people. You, dear Reader, now know how to add value through savvy and strategic social media to the organizations you join, the companies that employ you, to the careers of both your supervisors and those who work for you—everything you need, as that awesomely effective Vulcan educator of another era, Mr. Spock, put it, to "live long and prosper."

As I said when we began, you don't just want to survive, you want to prosper, and you want those around you to prosper. The social marketing era provides an enthusiastic and entrepreneurial individual with extraordinary opportunities to grow your career and company. And, to apply Homer once again: Ambition, community, and technology, is there anything they can't do!

(ENDNOTES)

1   The website is http://www.1to1media.com/contentchannels . Consider this a plug—I have no connection with this group, but have used their approaches and wisdom in class after class, taking their real-world view of the real-world view and bringing it to students. This has been my textbook for a number of years, and this is where I get to say "Thanks, guys and gals—you're doing so much good out there, bringing with it so many organizations to listen and interact with customers. Great reading, and each article an awesome piece of literature."

2   http://www.fastcompany.com/1811680/spotify-growing-by-8000-subscribers-per-day-more-than-netflix-sirius-xm

3   http://evolver.fm/2012/02/10/spotify-ceo-daniel-ek-talks-royalties-social-and-the-future/

4   Ibid.

ABOUT THE AUTHOR

DR. STUART H. SCHWARTZ is a professor of communication studies at Liberty University, specializing in advertising, direct and interactive marketing. He came to the school after a 25-year career as a senior executive with media and retail organizations. In addition, he is a popular columnist with one of the nation's leading conservative political/cultural websites, American Thinker. Schwartz has appeared on numerous radio talk shows, including the Dennis Miller Show, and has been mentioned by Rush Limbaugh. He received a B.A. from the University of Connecticut, an M.A. in Corporate & Political Communication from Fairfield University, and a Ph.D. in Communication from Temple University in Philadelphia. In addition, he holds an M.A. from Liberty Baptist Theological Seminary.

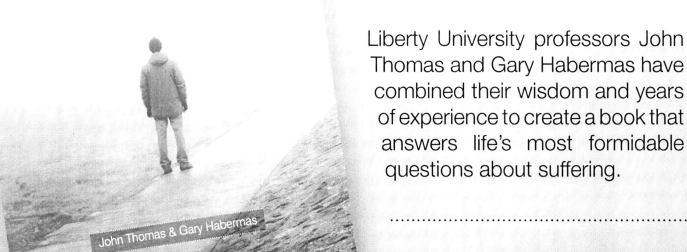

CPSIA information can be obtained at www.ICGtesting.com
Printed in the USA
LVOW09*1442010813

345831LV00011B/207/P